AI Agents Unleashed

The Next Big Thing

By
Quinn Bradford

AI Agents Unleashed

The Next Big Thing

Table of Contents

Introduction

We're on the brink of a new era, one where artificial intelligence (AI) agents are redefining the landscapes of technology, business, and everyday life. From the way we communicate to how industries function, AI agents are the harbingers of transformative change, ushering us into a future where intelligent automation aligns with human ingenuity. The path to understanding these dynamic agents begins here, as we embark on a journey of discovery and exploration.

At its core, the concept of AI agents is simple yet profound: machines that can perceive their environment, make decisions, and act autonomously to achieve specific goals. The very devices that once seemed like figments of our imagination are now top-of-mind conversations in boardrooms and living rooms alike. Tech enthusiasts, industry professionals, and curious minds find themselves on a collective quest to unravel the mysteries and potential these agents hold.

Yet, with every leap comes a challenge. As AI agents expand their reach, they bring to the forefront questions about ethics, security, and societal impact. It's imperative to understand not just what these agents can do, but how they integrate into the moral and regulatory frameworks of our world. While they promise efficiency and enhanced capabilities, they also ask us to reconsider privacy, bias, and the ever-blurring lines between man and machine.

Technologies like machine learning and natural language processing are just the tip of the iceberg. They're pivotal in making sense of colossal data volumes, enabling AI agents to learn, adapt, and develop problem-solving capabilities that mirror or even exceed human operators. Imagine a world where routine tasks are automated, allowing us to focus on creativity and complex problem-solving — this is where AI guides us.

The potential of AI agents isn't confined to a single domain. They're revolutionizing sectors as diverse as healthcare and education, providing personalized learning experiences and advanced diagnostic tools. In transport, they're driving us toward safer and more efficient systems, while in finance, they're crafting robust trading strategies and enhancing fraud detection efforts. Every domain touched by AI agents is a testament to their adaptability and vast potential.

However, with great power comes the need for great responsibility. As intelligent automation grows, so does the responsibility of molding it to benefit society. Addressing ethical challenges, such as data privacy and systemic bias, is as crucial as the technological advancements themselves. It's a balancing act of fostering innovation while safeguarding humanity's core values.

The narrative is as much about the future as it is about the present. As we look ahead, the promise of AI agents is intertwined with emerging technologies, transformations in the workforce, and new paradigms in human-machine collaboration. Future trends in AI agent development point to a continuum of innovation that reshapes global economies and redefines how individuals and societies interact.

In our exploration, we'll delve into how AI agents can further global integration, driving cross-cultural exchanges and influencing global economies. This raises considerations about education and skill transformation, as well as societal shifts and policy alterations needed to integrate AI-driven systems seamlessly.

This book aims to serve as a beacon for those navigating the vast seas of AI technologies, offering insights into not only where we're headed but how we can chart the course responsibly. By illuminating the transformative impact across sectors and contexts, we aspire to demystify these intelligent agents and encourage a dialogue that embraces their potential while acknowledging the challenges ahead.

The future with AI agents is an intricate tapestry of opportunities and obstacles. As we navigate this new world, our collective understanding and proactive adaptation will define how deeply these agents enrich our lives. It's an invitation for all of us to engage with these changes — to learn, to adapt, and to thrive in a world where AI aids and amplifies human potential.

The journey into the world of AI agents is more than a technological exploration; it's a story of growth, adaptation, and possibility. As we dive deeper into the nuances of these intelligent systems, we invite you to explore the present realities and future potential of AI agents with an open yet critical lens, embracing the intricate interplay of technology and humanity.

Chapter 1:
The Rise of AI Agents

The dawn of AI agents marks a transformative era, reminiscent of past technological revolutions but unparalleled in its potential to reshape our world. These digital entities, emerging from the confluence of advanced machine learning, robust computing power, and vast datasets, don't just perform tasks—they learn, adapt, and evolve. From improving efficiency in businesses to redefining human interaction through digital assistants, AI agents are everywhere. What was once the realm of science fiction is now integral to daily life, influencing industries from healthcare to entertainment. As we embark on this journey through the multifaceted landscape of AI agents, it's clear that they bring both profound opportunities and complex challenges. Understanding their ascent is crucial for anyone keen on navigating the future they herald, filled with both promise and ethical considerations that demand our attention.

Origins of AI Agents

The journey to today's AI agents begins with a confluence of visionaries, researchers, and technological milestones that have shaped what we now regard as artificial intelligence. The origins reach as far back as the mid-20th century, a period when the seeds of AI began to take root through speculative fiction and the ambition of forward-thinking scientists who dared to imagine machines capable of human-like thought and decision-making. This ambitious dream was underpinned by several key developments in mathematics, computer science, and cognitive psychology. Without the foundation they laid, AI agents as we know them today would likely remain the stuff of science fiction.

The 1950s witnessed the birth of the term "artificial intelligence," coined at the now-legendary Dartmouth Conference in 1956. This event brought together luminaries like John McCarthy, Marvin Minsky, Nathaniel Rochester, and Claude Shannon, who envisioned creating machines that could reason, learn, and even communicate. The concept of AI agents began to crystallize here. These initial gatherings and discussions provided a framework and a rallying point for researchers worldwide, marking the beginning of AI as a formal field of study.

In parallel with these philosophical and theoretical advances, the development of early computing technology paved the way for practical experimentation with AI. The creation of the first electronic computers in the 1940s and 1950s provided the raw computational power needed to test and refine early AI algorithms. The Electronic Numerical Integrator and Computer (ENIAC), one of the first general-purpose computers, demonstrated the potential of machines to handle complex calculations, laying the groundwork for future AI advancements.

As the 1960s progressed, significant strides were made in programming and algorithm development. Researchers began creating the first AI programs, such as the Logic Theorist and the General Problem Solver, which sought to mimic human problem-solving techniques. These endeavors marked the nascent stages of rule-based AI, which would later evolve into more sophisticated systems. The techniques developed during this period, although primitive by today's standards, were critical in shaping the architecture and function of contemporary AI agents.

Simultaneously, the philosophical dimensions of AI were explored and debated. Scholars and thinkers began to consider the implications of creating machines that could emulate human intelligence. Would they eventually surpass human capabilities? What ethical considerations would arise from their deployment? These questions framed the discourse surrounding AI, prompting deeper inquiry into the nature of intelligence itself and setting the stage for the ethical frameworks we continue to develop today.

By the 1970s, AI research had achieved several milestones, although not without setbacks, particularly in overstretched promises leading to periods known as "AI winters." This era saw the rise of expert systems, which were among the first forms of AI agents to find practical applications in industry and academia. These systems used a set of rules to simulate the decision-making ability of a human expert, and their success in specialized domains illustrated the potential of AI to transform industry practices, albeit within the constraints of the technological limits of the time.

Entering the 1980s and 1990s, AI saw a resurgence thanks to advances in computational power and the development of more robust algorithmic frameworks. Machine learning, a subfield of AI that allows systems to learn and improve from experience, began to gain prominence. While the concept was not entirely new, the ability to apply statistical methods to vast datasets brought unparalleled accuracy and functionality to AI agents. These advances laid the groundwork for the explosion of AI capabilities in the following decades.

Reinforcement learning emerged as another critical breakthrough during this period, providing a new paradigm in which agents learn to make decisions based on the rewards they receive for their actions. This approach brought AI closer to approximating human-like decision-making, emphasizing exploration and trial-and-error as a path to better performance. By focusing on outcomes rather than direct programming, reinforcement learning allowed AI agents to adapt to environments more dynamically and effectively.

The dawn of the 21st century marked a turning point in the trajectory of AI agents. With the exponential growth in data, driven by the internet and digital transformation across industries, AI agents became increasingly sophisticated and versatile. The confluence of vast data availability, increased computational power, and novel algorithms coalesced to create a favorable environment for AI advancement. Deep learning, in particular, revolutionized the capabilities of AI agents by harnessing neural networks to model complex patterns akin to human neural processing.

Concurrently, AI transitioned from a primarily academic pursuit to a driving force in global technology innovation. Tech giants like Google, Facebook, and IBM invested heavily in AI research, applying it to real-world problems in unprecedented ways. This commercialization of AI saw the rise of ubiquitous AI agents like chatbots, virtual assistants, and recommendation systems, integrating seamlessly into everyday life and transforming how we interact with technology.

Despite their growing presence and effectiveness, these AI agents also brought forth unique challenges and questions about their implications on privacy, ethical AI deployment, and societal impacts. The ongoing discourse now includes discussions on the responsibilities of developers and policy-makers to ensure AI agents are employed in ways that augment human experience while remaining cognizant of their potential downsides. This balance is crucial as AI continues to permeate all facets of life.

In retrospect, the origins of AI agents reveal a complex tapestry of discovery, innovation, and reflection. From the germ of an idea in the mid-20th century to the multifaceted systems of today, the development of AI agents demonstrates the tenacity and creativity inherent in technological progress. It underscores a shared ambition to push the boundaries of what machines can do and invites an ongoing dialogue about what the future holds. As we stand at the precipice of new possibilities, understanding this journey facilitates a deeper appreciation of the AI agents that have emerged and those yet to come.

Influential Technologies Behind AI Agents

The journey to creating intelligent AI agents is intertwined with the development and integration of multiple technologies. At the core, AI agents are products of advances in computing power, data availability, and algorithmic design. These advances provide the foundational blocks needed for AI agents to perform the complex tasks they tackle today. Let's dive into these influential technologies that have propelled AI agents from mere concepts to the sophisticated entities we interact with.

Perhaps one of the most profound influences on AI agents is the rise of **machine learning**. Machine learning, a subset of artificial intelligence, empowers AI agents to learn and adapt over time. By analyzing vast amounts of data, these agents can identify patterns and generate predictions without explicit instructions. This capability transforms a static system into one that evolves with new information, making AI agents more reliable and effective. Importantly, both supervised and unsupervised learning methods play a role in this adaptive learning process, enabling flexibility in approaching various tasks.

Another cornerstone technology is **natural language processing** (NLP). NLP bridges the gap between human language and computer understanding, allowing AI agents to interpret and respond to human communication effectively. This branch of AI delves into speech recognition, sentiment analysis, and language generation, powering the digital assistants we engage with daily. As NLP continues to evolve, it brings AI agents closer to achieving fluid, human-like interaction, which is critical for applications in customer service and personal assistance.

Deep learning, which builds upon machine learning principles, revolutionizes how AI agents process complex inputs such as images, sounds, and intricate datasets. By harnessing neural networks that mimic the human brain, deep learning enables AI agents to process multi-dimensional data efficiently. For instance, it allows for robust facial recognition systems and enhances AI's vision capabilities, crucial for autonomous vehicles and security systems. This technology's ability to handle unstructured data expands AI agents' usability across numerous sectors.

The role of **big data analytics** can't be overstated in the development of AI agents. With the proliferation of data generated by digital devices and sensors, AI agents require sophisticated analytics to process and analyze this information. By employing big data techniques, AI agents extract actionable insights, making them invaluable in fields ranging from finance to healthcare. Such insights drive decision-making processes, demonstrate patterns, and predict future trends, all vital for achieving operational efficiency and precision.

Another pivotal technology is the advancement in **robotics** and its integration with AI systems. Robotics enhances the physical manifestations of AI, transforming them from virtual constructs to tangible entities capable of interacting with the real world. AI-enabled robots can perform tasks in challenging environments unfit for humans, such as handling hazardous materials or exploring underwater ecosystems. By enabling such interactions, robotics extends the reach and impact of AI agents beyond the digital realm.

While these technologies are integral, one should not overlook the importance of **cloud computing**. Providing scalable resources and massive data storage solutions, cloud computing forms the backbone for many AI applications. It offers the computational power necessary for running complex AI algorithms and supports the development and deployment of AI agents at scale. This access to extensive computational resources democratizes AI development, allowing innovators and businesses to experiment with and refine AI models without the need for substantial on-premises infrastructure.

Supporting AI agents' communication and interaction capabilities, **internet of things** (IoT) technology links billions of devices worldwide. IoT devices serve as both data sources and points of interaction for AI agents, greatly expanding the latter's functionalities. In smart homes, healthcare monitoring, and smart city initiatives, IoT devices collect and transmit data, which AI agents analyze to offer real-time insights and services. This synergy enables personalized and responsive AI systems that seamlessly integrate into daily life and business operations.

Complementing these technologies, **blockchain technology** introduces an interesting dimension to AI agent development. It offers features of transparency, security, and immutability, potentially resolving issues around data integrity and trust in AI systems. Blockchain's decentralized nature aligns well with AI agents that require secure and verifiable data transactions, adding another layer of reliability to their deployments.

The convergence of these technologies doesn't occur in isolation; it's a collaborative evolution propelling AI agents towards greater autonomy and sophistication. By combining machine learning, NLP, and deep learning with cloud computing and IoT, AI agents become more versatile in handling a wide array of functions. The robustness of robotics and the security of blockchain further elevate the impact of

AI agents across industries, promising a future where human and AI collaboration is seamless and productive.

Looking ahead, the continuous refinement of these influential technologies holds exciting possibilities for AI agents. As computing capabilities grow, as algorithms become more advanced, and as our understanding of human-AI interaction deepens, AI agents are poised to push the boundaries of innovation. Their potential to address complex global challenges and improve quality of life inspires a future brimming with possibilities. While challenges remain, the technologies driving AI agents forward offer compelling evidence that their continued rise will be a transformative force in our society.

Chapter 2:
Core Technologies of AI Agents

As AI agents continue to evolve, their core technologies become increasingly pivotal, forming the foundation on which advanced capabilities are built. At the heart of these cutting-edge systems are complex algorithms that enable machines to learn, adapt, and make decisions with remarkable precision. Machine learning, a catalyst of modern AI, empowers agents to process vast amounts of data, recognizing patterns that surpass human perception. Alongside, natural language processing enables seamless communication between humans and AI, transforming how we interact with technology by interpreting and responding to human language in real-time. These technologies are not just enhancing the efficiency of AI agents but are fundamentally transforming industries and daily life, fostering a new era of intelligent automation that promises to redefine the boundaries of what's possible. In this evolving landscape, understanding the core technologies that drive AI agents offers a window into the future, challenging us to harness their potential thoughtfully and innovatively.

Machine Learning Fundamentals

Machine learning (ML) forms the backbone of modern AI agents, driving the capabilities that make these systems intelligent and adaptive. At its core, machine learning is the process by which computers are enabled to learn from data, identify patterns, and make decisions with minimal human intervention. This essential technology mimics the way humans learn, using algorithms to process vast amounts of data and improve over time. Let's delve into the foundational aspects of machine learning and explore its significance within the realm of AI agents.

The essence of machine learning lies in its ability to evolve. Unlike traditional software that relies on hard-coded rules, machine learning algorithms harness data to train models that can predict outcomes or classify information. One can think of it as providing a system with examples from which it can derive knowledge and insights. The more data it ingests, the more precise its predictions become, leading to higher efficiency and accuracy in tasks such as speech recognition, image analysis, and financial forecasting.

There are three primary types of machine learning: supervised learning, unsupervised learning, and reinforcement learning. Each type serves a unique purpose and is selected based on the task at hand. Supervised learning involves training a model on a labeled dataset, where the desired output is already known. This method is particularly effective for tasks like image classification or churn prediction, where historical data can guide future predictions. In contrast, unsupervised learning works with unlabeled data, identifying hidden patterns or intrinsic structures within the data. Techniques like clustering and dimensionality reduction fall under this category, playing a crucial role in customer segmentation or anomaly detection.

Reinforcement learning, the third type, is inspired by behavioral psychology and involves learning through trial and error. In this setup, an agent interacts with an environment and receives feedback in the form of rewards or penalties. Over time, the system learns strategies that maximize cumulative rewards. Reinforcement learning is instrumental in robotics, game playing, and real-time decision-making scenarios like traffic management or stock trading. It exemplifies the adaptability of machine learning, allowing systems to learn and develop strategies in dynamic and complex environments.

The magic of machine learning, however, is not merely in the algorithms but in the data that fuels these algorithms. Data is the lifeblood of machine learning models. With the exponential growth of big data, AI agents have become adept at turning raw information into actionable insight. However, this reliance on data also poses significant challenges. Data quality, privacy concerns, and the need for substantial computing resources are critical issues that researchers and practitioners must address to harness the full potential of machine learning.

Another core component that enhances the capabilities of machine learning is the neural network. Inspired by the human brain's structure, neural networks consist of interconnected layers of neurons, each performing a simple computation. When these neurons work together, they can solve complex problems like recognizing a face in an image or deciphering a spoken sentence. As one moves deeper into the layers of these networks, the system can capture more intricate patterns, enabling AI agents to understand and interpret the world in a sophisticated manner.

Deep learning, a subset of machine learning, leverages these deep neural networks to tackle tasks that were once considered intractable for machines. It is this very technology that has led to groundbreaking breakthroughs in image recognition, natural language processing, and autonomous vehicles. The deep layers allow AI to detect subtle differences and learn rich representations, making deep learning an indispensable part of modern AI systems.

While machine learning holds tremendous potential, it is not without its challenges. One of the most significant hurdles is the explainability of its models. As these models grow in complexity, understanding how they arrive at a particular decision can become opaque. This lack of transparency can be problematic, especially in sectors where decisions have far-reaching consequences, such as healthcare or law enforcement. Efforts are underway to develop interpretable models and methods to ensure that the decisions and predictions of AI agents align with human values and ethics.

Moreover, the ethical implications of machine learning can't be overstated. Bias in data can lead to biased outcomes, perpetuating existing inequalities in various domains. As machine learning systems often reflect the biases of the data they are trained on, there's a pressing need for ongoing scrutiny and improvement in how data is collected, processed, and deployed. Addressing these ethical challenges is crucial for ensuring that AI agents operate fairly and justly in their decision-making.

Despite these challenges, the transformative power of machine learning continues to propel AI agents to new heights. As these systems become smarter and more integrated into the fabric of our daily lives, they hold the promise of unlocking new horizons of automation and intelligence. From optimizing supply chains to enhancing personal digital assistants, the applications are as diverse as they are impactful.

As we continue to refine machine learning techniques, the potential for AI agents to revolutionize industries and societal norms remains vast. Innovations in transfer learning, federated learning, and edge computing highlight the dynamic evolution of the field, opening doors to applications that were once purely speculative. Each leap forward brings us closer to an era where AI agents can learn and adapt autonomously, paving the way for a future that, while not without challenges, is ripe with opportunity.

In conclusion, machine learning stands as a cornerstone of AI agents, imbuing them with the capability to learn, adapt, and excel across a myriad of applications. Its fundamental principles provide a foundation upon which cutting-edge AI technologies are built. As we delve deeper into this journey, machine learning will undoubtedly continue to shape the future of how we interact with the world through intelligent agents.

Natural Language Processing in AI Agents

Natural Language Processing, commonly referred to as NLP, stands at the heart of making AI agents genuinely intelligent and interactive. It's the bridge that connects human communication with machine comprehension. In the realm of AI, understanding and generating human language is more than just a technical challenge—it's a revolutionary leap toward seamless integration of machines in daily life. Imagine communicating with robots, digital assistants, and smart devices just as you would with another person. This is what NLP aims to achieve.

The complexity of NLP lies in its requirement to comprehend the nuances, contexts, and subtleties of human language. Language is inherently ambiguous and often open to interpretation, making it a formidable task for machines to grasp its full meaning accurately. Words have varied meanings based on context, tone can alter the messages' essence, and idiomatic expressions can confuse even proficient human speakers. For AI agents, overcoming these challenges is essential to transform simple dialogue into a sophisticated interaction.

Today's AI agents owe much of their conversational prowess to advancements in NLP technologies. At its core, NLP involves several crucial processes, including tokenization, parsing, semantic understanding, and language generation. Tokenization breaks down text into meaningful units. Parsing then analyzes grammatical structures, while semantic understanding ensures the agent grasps the significance behind the words. Finally, language generation enables AI agents to respond coherently and contextually in human-like ways.

Natural Language Processing has seen rapid evolution with the advent of deep learning techniques. Models like Transformers, particularly BERT and GPT, have elevated NLP capabilities by leveraging large datasets to learn language patterns deeply and contextually. These models are designed to understand and generate text, enabling AI agents to engage in more fluid and natural-sounding conversations. The innovation lies not just in processing massive volumes of data but in discerning the subtle connections within the language, allowing AI to mimic human-like dialogue. The result is an enhanced user experience that feels intuitive and, at times, almost indistinguishable from speaking with a human.

AI agents apply NLP in various applications across different sectors. In customer service, chatbots use NLP to assist consumers, answering queries, solving issues, and providing recommendations with increasing efficiency. These AI agents reduce response times and improve customer satisfaction by offering consistent support round the clock. In healthcare, NLP helps extract meaningful insights from clinical records, aiding in diagnosis and patient management. By processing vast amounts of textual data, AI agents can identify patterns and provide valuable information to medical professionals, contributing to better patient outcomes.

One of the most transformational aspects of NLP in AI agents is its role in democratizing access to information and services. Through language translation tools, AI agents break down language barriers, fostering inclusivity and accessibility in global communication. This capability empowers individuals from diverse linguistic backgrounds to interact seamlessly, driving cross-cultural exchanges and enhancing collaboration on a global scale.

However, NLP isn't without its challenges. One major hurdle is managing the inherent biases present in language data. Since AI models learn from data reflecting social norms, prejudices can inadvertently be perpetuated. Bias in training data can lead to skewed AI decision-making, raising significant ethical concerns. Thus, researchers and developers must diligently work on scrutinizing and refining datasets to ensure fair and unbiased language processing.

The future of NLP in AI agents looks promising as research continues to push the boundaries of what's possible. Developments in contextual understanding are paving the way for even more profound applications, such as interactive storytelling, emotion recognition, and psychological profiling. These advancements will further deepen our interactions with AI agents, making them indispensable companions in both personal and professional domains.

Moreover, as NLP technologies evolve, they contribute significantly to the progress of AI as a whole. By enhancing our ability to communicate with machines, NLP fosters greater integration of AI in various aspects of society, from simplifying mundane tasks in smart homes to enriching educational experiences through personalized learning. The continuous refinement of NLP technologies will undoubtedly spearhead the march towards sophisticated, empathetic, and utterly human-like AI agents.

In conclusion, Natural Language Processing is not just a component of AI technology—it's a cornerstone in creating agents that understand us, anticipate our needs, and respond naturally. By bridging the gap between human language and machine communication, NLP enriches our interactions with AI, heralding a future where intelligent agents are not only tools but partners in our journey through an increasingly digital world.

Chapter 3:
Key Types of AI Agents

As we delve deeper into the landscape of artificial intelligence, a critical focus shifts to the various types of AI agents that have emerged, each possessing distinct roles and functionalities. At the forefront are *digital assistants*, revolutionizing the way we interact with technology by adding layers of convenience and efficiency to our daily routines. Their seamless integration into our lives showcases how conversational AI and proactive aid can fundamentally transform user interaction. Another type worth noting are *autonomous agents*, pivotal in industries like manufacturing and logistics, where they enhance productivity by performing tasks that require minute precision and timeliness. These agents, leveraging real-time data and complex algorithms, are reshaping operational processes across sectors. Together, these varied AI agents offer a glimpse into a future that promises increased automation and smarter environments, challenging us to imagine the endless possibilities they hold for an intertwined world of humans and machines.

Digital Assistants: Transforming Interaction

Digital assistants have rapidly become the cornerstone of modern interaction, fundamentally altering how we engage with technology and each other. From the moment we wake up, these AI-driven marvels seamlessly integrate into our daily routines. Their ability to understand natural language and perform tasks autonomously has revolutionized the expectations we hold for interaction, setting a new standard for convenience and efficiency.

At the heart of this transformation lies the simplicity of voice command. Gone are the days when interacting with machines required navigating complex interfaces or memorizing commands. Today, a mere utterance suffices—"switch on the lights" or "what's the weather like?"—and digital assistants spring into action. The immediacy and intuitiveness of this interaction reflect a giant leap towards making technology accessible to all, including those who might have found traditional computing daunting.

Digital assistants like Amazon's Alexa, Apple's Siri, and Google's Assistant are not just tools for convenience; they're companions that grasp context, learn from interactions, and anticipate needs. They are continuously refining their capabilities, showcasing how advancements in natural language processing empower these agents to better understand and predict user intent. This evolution in understanding nuances, such as tone and context, elevates the interaction from a basic query-response dynamic to a more conversational and personalized experience.

However, this transformation goes beyond mere convenience. Digital assistants are a crucial gateway to smart home ecosystems. By serving as the control hub, they facilitate the symbiotic operation of multiple smart devices—from thermostats and security cameras to kitchen appliances—ushering in an era of unprecedented home automation. These agents allow users to focus on more important tasks while delegating mundane responsibilities, enhancing productivity and overall quality of life.

Beyond the home, digital assistants are also revolutionizing the workplace. Businesses are integrating these AI agents into their processes to streamline meetings, manage schedules, and provide insights from data. Employees find themselves engaging not only with coworkers but with tireless digital counterparts capable of mining information, synthesizing reports, and even participating in decision-making processes. The fluid and dynamic interaction facilitated by digital assistants is reshaping how information flows within and outside organizations, breaking down silos and fostering collaboration.

This transformation raises profound questions and challenges. How do we ensure the privacy of users when interactions are becoming increasingly ubiquitous? The ability of digital assistants to process vast amounts of personal data necessitates robust data protection frameworks. Users must trust that their assistants safeguard sensitive information without misuse. This trust is built through transparency in AI operations and stringent regulatory oversight, ensuring these intelligent agents are not only beneficial but also secure.

Furthermore, the cultural and ethical implications of these interactions cannot be ignored. As digital assistants become fixtures in daily life, they influence language, behavior, and expectations. They can be both educators and inadvertent perpetuators of bias. The design and programming of these agents thus bear a responsibility—developers must prioritize inclusive language models and diverse datasets to ensure equitable interactions for all users, regardless of their background or language proficiency.

The rise of digital assistants also reflects a broader trend in human-computer interaction: the blurring line between technology and human intuition. As AI systems learn to predict our preferences and preemptively respond to needs, they echo a deeper understanding akin to human empathy. This anticipatory service heightens the user experience, evoking a sense of personalized interaction that transcends mere functional utility.

Looking to the future, the potential for digital assistants to further transform interactions is vast. As machine learning and AI technologies continue to evolve, these agents will only become smarter, more intuitive, and more human-like in their interactions. Multimodal interfaces, integrating voice, gesture, and visual recognition, promise an even richer tapestry of interaction possibilities. We stand on the cusp of an era where these AI agents become an extension of our cognitive processes, seamlessly interfacing with the digital world with unprecedented sophistication.

In conclusion, digital assistants are not just changing how we interact with technology; they are redefining the fabric of human interaction itself. They are gateways to a future where technology serves us more intuitively and insightfully than ever before, paving the way for experiences that are as natural as they are transformative. As we embrace these changes, we must navigate the challenges they present with responsibility and foresight, ensuring a symbiotic relationship between humanity and technology.

Autonomous Agents in Industry

Within the expansive landscape of artificial intelligence, autonomous agents are not just emerging as technological wonders but as essential cogs in the machinery of modern industry. Defined by their ability to perform tasks independently, these agents are programmed with a specific goal in mind and operate without human intervention. Their deployment marks a significant shift in how industries achieve efficiency and innovation, adapting to a context where decision-making speed and accuracy are paramount. Autonomous agents can potentially transform entire sectors by autonomously adapting, learning, and optimizing processes in real-time.

Manufacturing is undeniably at the forefront of this revolution. Thanks to autonomous agents, production lines are evolving from rigid, assembly-line structures into dynamic entities capable of adapting to varying demands and conditions. Cutting-edge factories now employ AI-driven robotics that repair themselves, optimizing workflows to minimize downtime. These agents enable plants to maintain output quality while reducing waste and energy consumption significantly. The learning algorithms enhance not just productivity but sustainability, a key driving force in today's industries focused on reducing their carbon footprint.

In logistics, the efficiency and reliability of supply chains have long been hindered by unforeseen disruptions such as weather events or geopolitical tensions. Autonomous agents are game-changers here, providing predictive analytics and making real-time decisions to reroute goods, anticipate demand fluctuations, and optimize delivery routes. This not only saves time and costs but also ensures a higher level of service quality. The deployment of autonomous delivery units, from drones to driverless trucks, demonstrates a tangible shift toward seamless logistics operations, reducing human error and increasing safety on the roads.

The industrial sector's utilization of autonomous agents extends into quality control. Machines equipped with advanced vision systems can detect imperfections more precisely than the human eye, ensuring that only superior products reach customers. These agents tirelessly inspect, assess, and sort products, maintaining consistency and quality at levels previously unattainable. By constantly learning from their assessments, these systems can even predict future manufacturing anomalies, facilitating maintenance schedules that prevent costly downtimes.

In energy industries, the deployment of autonomous agents is leading to profound changes. Smart grids powered by AI can balance production and consumption demand with remarkable precision, predicting peak times and distributing energy appropriately. Autonomous wind and solar farms make micro-adjustments to capture more energy efficiently, compensating for variable environmental conditions. This level of optimization guarantees stability in energy availability, aligning production closely with consumption needs.

The realm of finance is another arena where autonomous agents are crafting transformative impact. Financial systems harness autonomous agents to monitor market trends, execute trades within nanoseconds, and identify fraud patterns that elude traditional systems. By acting independently and learning from vast datasets, these agents reduce human bias and error, driving financial markets toward more stable and transparent from systemic discrepancies.

In all these applications, the advantages are clear: reduced operational costs, increased efficiency, and minimized errors. Yet, these benefits come with challenges that industries must navigate. One such challenge is the integration of autonomous systems with existing infrastructure, which often involves substantial investment and a redesign of processes. Furthermore, industries need to address the human aspect of this transformation—ensuring that their workforce is equipped with the necessary skills to collaborate with these intelligent systems.

The ethical implications of autonomous agents in industry cannot be overlooked. As decision-making becomes increasingly reliant on AI, the accountability for decisions made by autonomous systems remains a crucial subject. Industries must ensure transparency in AI decision-making processes, preserving trust among stakeholders. There is also the critical task of programming these agents with ethical considerations that mirror human values and societal norms, preventing any unintended consequences.

Innovation in autonomous agents is expanding rapidly, with industries like agriculture and construction now adopting AI-driven technologies. Precision farming techniques, guided by autonomous drones and soil sensors, are revolutionizing agricultural productivity by optimizing planting patterns and pest control measures. Meanwhile, autonomous construction equipment can operate in the most hostile environments, enhancing safety and productivity by performing tasks that are dangerous for humans.

As industries continue to leverage these technologies, the potential impact of autonomous agents becomes vast and varied. The confluence of advancements in machine learning, robotics, and IoT (Internet of Things) is creating industrial ecosystems that are interconnected and intelligent. By harnessing data, these agents are not just components of an industrial system; they become its nerves and brains, facilitating decisions that drive progress and efficiency.

In summary, the surge of autonomous agents in the industry is more than an upgrade of current systems; it's a paradigm shift that redefines operational capabilities and strategic planning. This transformation requires industries to embrace changes and foster innovation, ensuring that these systems align with corporate values and societal needs. Balancing the benefits with ethical practices will enable industries to not just survive in this automated era, but to thrive.

Looking forward, the success of this integration hinges on collaboration between technologists, policymakers, and industry leaders, converging on frameworks that guide ethical AI integration and workforce transformation. It's a dynamic and challenging journey, but one brimming with opportunities, promising a future where autonomous systems in industry pave the way for unprecedented efficiency and innovation.

Chapter 4:
AI Agents in the Business World

In today's business landscape, AI agents are reshaping industries with unprecedented speed and agility. They're no longer just futuristic concepts; they're integral components of operations across the globe. These intelligent systems are revolutionizing how companies streamline their processes, making operations more efficient and minimizing costs. By analyzing vast quantities of data in real-time, AI agents offer insights that are both predictive and prescriptive, allowing for more informed decision-making. Moreover, they enhance customer experiences by providing personalized interactions that go beyond traditional customer service. Businesses are now equipped to offer 24/7 support, understanding customer needs and preferences with ever-increasing accuracy. The synergy between AI agents and business is fostering innovation, driving competitive advantage and redefining what's possible in the corporate world.

Streamlining Operations with AI Agents

In the ever-evolving landscape of the business world, AI agents have emerged as pivotal tools in streamlining operations across various industries. These intelligent systems are designed to automate routine tasks, optimize processes, and drive efficiency, allowing organizations to focus on strategic growth and innovation. By integrating into existing workflows, AI agents can help businesses cut costs, reduce errors, and enhance productivity.

One of the primary benefits of deploying AI agents in operations is their ability to handle repetitive tasks with minimal supervision. Duties that once required significant human intervention, such as data entry, inventory management, and scheduling, can now be automated to run seamlessly in the background. This approach not only frees up valuable human resources but also significantly minimizes the risk of human error, ensuring that operations are conducted with precision and consistency.

Consider the example of supply chain management, a complex network that requires careful coordination. AI agents can analyze vast amounts of data to predict demand, manage inventory, and coordinate logistics, ensuring that products are delivered efficiently and cost-effectively. By optimizing these processes, businesses can reduce waste, enhance customer satisfaction, and maintain a competitive edge in the market.

In manufacturing, AI agents are transforming operations by introducing smart automation and robotics. These agents are capable of monitoring production lines, identifying bottlenecks, and suggesting improvements to enhance efficiency. For instance, they can adjust machinery settings in real-time to optimize production and reduce downtime, making operations more agile and responsive to market demands.

Moreover, AI agents can provide real-time insights and analytics, enabling businesses to make data-driven decisions. By processing large datasets quickly and accurately, these agents offer actionable insights that can lead to improved strategy formulation and resource allocation. Companies can better anticipate market trends, identify new opportunities, and make informed decisions that propel growth.

Another significant advantage of AI agents is their role in enhancing communication within organizations. These systems can manage internal messages, automate emails, and organize meetings, streamlining communication and ensuring that information flows smoothly across departments. By doing so, they minimize misunderstandings and enhance collaboration, ultimately fostering a more cohesive working environment.

AI agents can also contribute to quality assurance, monitoring product standards and flagging any discrepancies in real-time. This proactive approach allows businesses to maintain high-quality products, reducing returns and increasing customer confidence. Additionally, AI-driven predictive maintenance can anticipate equipment failures, prompting timely interventions that avoid costly disruptions.

The financial operations of a business can also benefit from AI agents. They can automate tasks such as invoice processing, payroll calculations, and financial reporting. These agents help ensure compliance, improve accuracy, and provide valuable insights into financial performance. With AI handling these intricate tasks, businesses can maintain a robust financial health while focusing on strategic growth.

Personnel management is another area where AI agents are making a significant impact. They can automate various HR functions, from recruitment to employee onboarding, training, and performance evaluation. By continually analyzing employee data, AI agents offer valuable insights that can improve employee satisfaction and retention rates.

Of course, while the advantages of AI agents in streamlining operations are manifold, implementing these systems does not come without challenges. Businesses must tackle concerns around data privacy, system integration, and the need for significant initial investments. Furthermore, fostering a culture that embraces AI-driven change is crucial for realizing the full potential of these technologies.

In today's rapidly changing business environment, those who successfully integrate AI agents into their operations stand to gain a significant advantage. By capitalizing on the speed, efficiency, and analytical power of AI, organizations can streamline operations in ways that were once considered unimaginable.

In conclusion, AI agents are revolutionizing business operations across the board, transforming how companies manage daily tasks and make strategic decisions. Through intelligent automation, enhanced efficiency, and insightful data analytics, AI agents are empowering businesses to streamline their processes like never before. These transformations are just the beginning, as AI technology continues to evolve, promising even more innovative applications in the future.

Enhancing Customer Experience through Intelligent Agents

The customer experience has undergone a radical transformation with the advent of intelligent agents. These AI-powered assistants are redefining how businesses interact with their customers, offering a level of personalization and responsiveness that was once unimaginable. In an era where consumers have higher expectations than ever before, intelligent agents are bridging the gap between what companies offer and what customers desire. They achieve this by providing timely information, resolving queries swiftly, and tailoring user experiences to individual preferences.

One of the most significant impacts of intelligent agents in enhancing customer experience is their ability to offer 24/7 support. Unlike traditional customer service centers that operate within specific hours, AI agents never sleep. They're always available to answer questions, troubleshoot problems, and provide support. This constant availability reduces wait times and ensures that customers can get the help they need exactly when they need it. By doing so, businesses not only improve customer satisfaction but also foster loyalty and trust.

AI agents excel in personalizing customer interactions, which is crucial for creating memorable experiences. Through sophisticated data analytics and machine learning algorithms, these agents can analyze previous interactions, purchase histories, and browsing behaviors to offer tailored recommendations and solutions. This level of customization makes customers feel valued and understood, prompting positive feelings towards a brand. For instance, e-commerce platforms utilize AI agents to recommend products that align with a user's past purchases and browsing history, significantly enhancing the shopping experience.

Another advantage of using intelligent agents is their capacity to handle large volumes of queries simultaneously. Traditional customer service models can become overwhelmed during peak times, leading to frustrating wait times and hurried interactions. AI agents, on the other hand, can process multiple inquiries at once without compromising on response quality. This scalability allows businesses to efficiently manage demand fluctuations, ensuring consistent service quality regardless of the time of day or the volume of incoming questions.

AI-driven chatbots and voice assistants are among the most common forms of intelligent agents used for enhancing customer experience. These technologies leverage natural language processing to understand and interpret human speech, allowing for seamless and intuitive customer interactions. By mimicking human conversation, they provide a human-like touch to digital interactions, which can be particularly comforting for customers seeking assistance. This technological sophistication instills confidence in users, making interactions less transactional and more conversational.

Intelligent agents are also instrumental in gathering customer feedback and insights. They can engage users in brief surveys or solicit feedback after an interaction, providing businesses with valuable information on customer preferences, pain points, and satisfaction levels. This data can then be analyzed to identify trends, improve services, and develop new products tailored to meet customer needs more effectively. By closing the feedback loop in real time, companies can rapidly adjust their strategies to keep pace with evolving consumer demands.

Furthermore, intelligent agents contribute to enhancing customer experience by integrating seamlessly with other systems and technologies. For example, linking AI agents with CRM (Customer Relationship Management) systems allows businesses to leverage customer data more effectively, ensuring that all interactions are informed by the latest user information. This integration leads to greater coherence and continuity across customer touchpoints, reducing the need for customers to repeat information as they transition between different service channels.

The benefits of intelligent agents also extend to optimizing the operational aspect of customer experience management. By automating routine inquiries and tasks, AI agents free human employees to focus on more complex and high-value activities. This shift not only enhances employee productivity but also enables them to engage in deeper, more meaningful interactions with customers who require human intervention. The symbiotic relationship between AI agents and human employees results in a more efficient and customer-centric operation.

Nevertheless, integrating intelligent agents into customer service strategies is not without challenges. Ensuring these agents have access to up-to-date information is crucial for maintaining their effectiveness. Businesses must invest in robust infrastructure that supports real-time data access and processing. Moreover, continuous training of AI agents is necessary to keep them apprised of the latest products, services, and policies. This dynamic adaptability is essential for providing accurate and relevant customer support.

Organizations must also address the ethical implications of using intelligent agents in customer interactions. Maintaining transparency about the use of AI and safeguarding customer data are paramount. Businesses must clearly communicate when a customer is interacting with an AI agent and provide options for escalating complex issues to human representatives. By prioritizing transparency and ethical considerations, companies can mitigate potential privacy concerns and build trust with their users.

In the end, intelligent agents are driving a customer service revolution. They're fundamentally reshaping how businesses engage with customers by providing faster, more personalized, and more insightful interactions. As technology continues to evolve, the capabilities of these AI agents will only increase, offering even greater opportunities for enhancing customer experience. Businesses that embrace these intelligent tools are well-positioned to lead in a competitive landscape, delivering unparalleled service quality that meets the demands of the modern consumer.

Chapter 5:
AI Agents in Healthcare

AI agents are reshaping the landscape of healthcare, offering transformative potential that could redefine patient care and medical practice. With their unparalleled ability to analyze vast amounts of data quickly, these intelligent systems enhance diagnostic accuracy and assist medical professionals in providing personalized treatment plans. By balancing complex algorithms and human intuition, AI agents empower clinicians to make data-driven decisions, improving outcomes and efficiency. The integration of AI in healthcare doesn't merely streamline operations; it heralds a new era of proactive and preventive care. As technology continues to evolve, AI's role will expand, addressing challenges and unlocking possibilities, ultimately crafting a future where technology and humanity collaborate to save and improve lives.

Diagnostic Tools Powered by AI

In the realm of healthcare, the integration of artificial intelligence (AI) in diagnostic tools represents a seismic shift, one that's reshaping how we approach disease detection and patient care. While traditional diagnostic methods have served us well, they often require considerable time and resources to deliver precise results. AI's emergence infuses these processes with unprecedented efficiency, speed, and accuracy, driving a transformation that's becoming impossible to ignore.

A primary advantage of AI in diagnostics is its ability to process and analyze vast databases of medical images more rapidly than human capability. Consider radiology, where AI algorithms are now capable of sifting through countless MRI, CT scan, and X-ray images to detect anomalies with an accuracy that rivals experienced clinicians. These AI-driven systems have learned to identify patterns and irregularities that might elude even the most trained eyes, due to their capacity to ingest and interpret huge volumes of data promptly. This capability is not about replacing radiologists but augmenting their abilities, enabling them to focus on complex cases that require a human touch.

The potential of AI in predictive diagnostics is another area gaining momentum. Predictive analytics harnesses machine learning models to foresee future health issues by analyzing historical patient data. By identifying risk factors and trends, AI can offer early warnings for diseases like diabetes, heart disease, and cancer long before symptoms manifest. This predictive power is crucial in preventive medicine, where the goal is to intervene before the onset of a potentially life-threatening condition. For clinicians, this means more proactive healthcare and personalized treatment plans tailored to individual risk profiles.

AI's role doesn't stop at imaging and prediction. Consider the realm of pathology, where AI systems assist in examining tissue samples under the digital microscope. Traditionally, this process is painstakingly slow and labor-intensive. AI-powered systems streamline the evaluation by highlighting significant features and automating quantification tasks, cutting down on human error and liberating pathologists to make more informed decisions. Consequently, the diagnostic cycle is accelerated, benefiting patients who get their results sooner.

Moreover, with the digital transformation of healthcare data through electronic medical records (EMR), AI systems have a treasure trove of information to tap into. By mining EMRs, AI can generate insights about patient histories, treatment outcomes, and even population health trends. This wealth of information helps refine diagnostic tools further, supporting decision-making that reflects the most current medical knowledge and practice. Real-time data analysis means that every new piece of information can immediately enhance predictive accuracy and decision support.

Ethical considerations are paramount in deploying AI diagnostic tools. There's a need for transparency in AI algorithms to ensure reliability and trust. Stakeholders, including patients and practitioners, must understand how AI arrived at its conclusions. This transparency builds trust in the diagnostics provided, a crucial aspect of healthcare where human lives depend on these decisions. Furthermore, AI systems require extensive, unbiased training datasets to avoid perpetuating existing disparities within healthcare systems.

Achieving greater automation in diagnostics through AI also raises regulatory challenges. These advanced tools must undergo rigorous testing to meet the standards set by health authorities before becoming part of clinical workflows. AI developers collaborate with regulatory bodies to establish frameworks that ensure safety and efficacy while fostering innovation. This balancing act is critical, as the pace of technological advancement often surpasses existing regulatory frameworks.

Despite these challenges, the incentives to expand AI's diagnostic role are significant. Reduced diagnostic times lead to faster treatment initiation, which is particularly critical in diseases where early intervention determines patient outcomes. The healthcare system as a whole benefits from efficiencies, freeing up resources and reducing costs associated with prolonged hospital stays and treatments initiated too late.

Inspiration for AI's future in healthcare diagnostics continues to draw from interdisciplinary collaborations, melding insights from computer science, medicine, and cognitive psychology. Bringing AI into the heart of healthcare requires not only technological advancement but also a cultural shift where technology and humanity coexist symbiotically. This shift mandates an alliance reinforcing the strengths of each, where AI's analytical acumen complements the nuanced understanding of human biology unique to healthcare professionals.

It's fascinating to ponder the future trajectories of AI in diagnostics. As AI systems become more sophisticated, we're approaching scenarios where they could autonomously monitor patients post-diagnosis, adjusting treatments based on real-time data and responses. This heralds a level of personalized care that's anticipatory versus reactive, driven by continuous learning and adaptations.

The transformation that AI diagnostics promises is not just technological but deeply human. It underscores medicine's core ethos: deliver care that is smarter, faster, and ultimately more empathetic. As AI navigates intricacies that were once believed to be the exclusive domain of human intellect, a harmonious future where human compassion and machine precision create a potent force for good in healthcare becomes more than a vision—it becomes reality.

AI Agents Assisting Medical Professionals

In the rapidly evolving landscape of healthcare, AI agents are emerging as indispensable allies for medical professionals, transforming the way they diagnose, treat, and manage patient care. These AI-powered tools are designed to enhance the capabilities of healthcare providers, offering insights that were once unimaginable. By integrating seamlessly into existing medical workflows, AI agents are not only streamlining processes but also ushering in a new era of precision medicine.

One of the most notable contributions of AI agents in healthcare is their ability to analyze vast amounts of data quickly and accurately. Medical professionals are often inundated with data, ranging from electronic health records to complex imaging results. AI agents can sift through this information, identifying patterns and abnormalities that might be missed by the human eye. This capability is particularly transformative in fields like radiology and pathology, where AI agents can assist in detecting anomalies that signal early stages of diseases.

Take, for example, the use of AI in imaging diagnostics. AI agents equipped with machine learning algorithms can analyze CT and MRI scans at an unprecedented speed, identifying potential issues faster than traditional methods. This not only accelerates the diagnostic process but also improves accuracy, reducing the chances of misdiagnosis. Consequently, patients receive timely and more effective treatments, improving outcomes and quality of life.

Beyond diagnostics, AI agents are also revolutionizing treatment planning. Personalized treatment regimens tailored to individual patient profiles can be created using insights generated by AI systems. These agents consider a myriad of factors, including genetic information, lifestyle, and even environmental influences, to suggest optimal treatment pathways. This level of personalization was previously unattainable and is now within reach, thanks to the computational prowess of AI.

Moreover, AI agents are proving invaluable in managing patient care. Virtual health assistants, powered by natural language processing, help manage patient inquiries, schedule appointments, and even provide reminders for medication intake. These agents ensure that patients remain engaged with their care plans, improving adherence and, ultimately, health outcomes. They also provide immense relief to healthcare workers by taking over administrative tasks that, while critical, consume considerable time and resources.

AI's role in drug discovery can't be understated, either. Traditional drug development is a lengthy and costly process, often taking over a decade to bring a new drug to market. AI agents are changing this by predicting how different molecules will interact with specific diseases, drastically shortening the time required for early-stage drug discovery. This capability accelerates the development of new therapies, especially crucial for combating rapidly evolving diseases and conditions.

AI agents also have a profound impact on telemedicine. As remote consultations become more common, especially in rural or underserved areas, AI-powered tools ensure that physicians have the information they need at their fingertips. These agents can pre-analyze patient data, suggesting potential areas of concern before consultations. This enables doctors to make informed decisions even in the absence of a physical examination, broadening access to quality healthcare.

Despite the immense benefits, the integration of AI agents in healthcare comes with its share of challenges. Data privacy remains a significant concern, as sensitive health information must be protected against breaches and misuse. Trust in AI-driven decisions is another critical issue, requiring transparency in how these systems arrive at their conclusions. Medical professionals must be equipped with a clear understanding of AI processes to make informed decisions and reassure patients.

The ethical considerations surrounding AI in healthcare also require attention. While AI agents can greatly assist in decision-making, human oversight remains essential. Ensuring that AI complements rather than supplants human judgement is key to ethically sound practice. Moreover, addressing biases inherent in AI systems is imperative to prevent skewed outcomes that could adversely affect patient care.

Looking forward, the potential of AI agents assisting medical professionals is immense and continues to grow. As these systems learn and evolve, their role will expand into areas like genomics and lifestyle medicine. AI agents are on the cusp of triggering a paradigm shift in healthcare, one that offers the promise of more personalized, accurate, and accessible care for all.

This transformative journey, however, requires a collaborative effort from technologists, policymakers, and healthcare professionals to harness AI's potential responsibly and ethically. By bridging the gap between technology and compassionate care, AI agents stand to revolutionize how we perceive and receive healthcare, heralding a future where medical professionals are empowered like never before.

Through this symbiotic relationship between AI and human expertise, the future of healthcare looks not only more efficient, but also more humane, as each advancement stands to fundamentally enhance the way we approach medicine, ensuring that the well-being of patients remains paramount.

Chapter 6:
Revolutionizing Education
with AI Agents

In the realm of education, AI agents stand poised to redefine how learners of all ages access and absorb information, turning conventional education models on their heads. By tailoring educational content to individual needs, AI offers personalized learning experiences that were once a distant dream. Imagine an intelligent tutoring system that adapts in real-time to a student's learning style, identifying gaps in knowledge while providing immediate feedback and resources to address them. These AI-driven systems foster a dynamic learning environment where students are empowered to take charge of their educational journey, leading to greater engagement and improved outcomes. As we continue to embrace this shift, educators are also finding that AI can handle administrative tasks, allowing them more time to focus on the human aspects of teaching that technology can't replicate. This synergy of human and artificial intelligence in education is setting the stage for an era of unprecedented educational opportunities and innovation.

Personalized Learning Experiences

In the realm of education, the application of artificial intelligence is nothing short of transformative. By leveraging AI to cater to the individual needs of learners, education is moving away from the traditional one-size-fits-all model. Personalized learning experiences are at the forefront of this revolution, offering tailored instruction that values the unique pace, interests, and learning style of each student.

AI agents have the remarkable ability to analyze a student's learning history, identify strengths and weaknesses, and adapt content accordingly. This process begins with the collection and analysis of vast amounts of student data—everything from quiz scores to the time spent on particular tasks. With these insights, AI can craft a curriculum that provides the right challenges at the right moments, thereby maximizing engagement and retention.

Personalized learning often involves the integration of AI-powered platforms that offer adaptive learning paths. These platforms adjust the difficulty of tasks based on real-time feedback, facilitating an environment where students can advance at their own pace. Such systems not only encourage self-paced learning but also reduce the frustration and pressure often associated with keeping up with peers.

Consider a mathematics student struggling with algebraic concepts. An AI agent, observing this difficulty, might redirect the student to supplementary resources or adjust the difficulty of subsequent exercises. This continuous feedback loop helps in creating a supportive learning atmosphere, where students have the freedom to explore subjects without the fear of failure.

The scope of such personalized experiences extends beyond academic subjects to include the development of critical thinking, problem-solving, and creativity. AI-driven platforms can incorporate gamified learning scenarios that make education fun and rewarding, encouraging students to apply their knowledge in creative ways. This kind of learning activates higher cognitive functions and encourages deep understanding, crucial for tackling real-world challenges.

Moreover, personalized learning experiences powered by AI aren't limited to any specific age group or educational sector. From primary education to lifelong learning, AI agents can provide value and customization at every stage. In corporate settings, AI can tailor professional development programs, ensuring that employees receive training that's both relevant and timely.

For teachers and educators, AI provides the ability to manage the increasingly diverse needs of students more effectively. Real-time analytics offer unprecedented visibility into student progress, allowing educators to make informed interventions where necessary. Also, by automating routine assessments, AI frees up valuable time for teachers to focus on relational aspects of education, such as mentoring and guidance, which are irreplaceable by technology.

In addition to supporting instructors, AI agents can serve as virtual tutors, accessible to students when teachers may not be available. These AI tutors can answer questions, clarify doubts, or provide practice exercises, all while maintaining a personalized approach that evolves with the student's learning journey. This kind of round-the-clock assistance significantly boosts confidence and autonomy among learners.

Access to personalized learning through AI can also play a crucial role in leveling the educational playing field. For students in underserved or remote areas, AI can bring high-quality education that might otherwise be unavailable. By minimizing educational disparities, AI-driven learning systems contribute positively towards reducing socioeconomic inequities within society.

Nonetheless, the deployment of personalized learning experiences powered by AI does not come without challenges. Concerns regarding data privacy and the ethical use of student information are significant. It's imperative to navigate these issues carefully to ensure that personalized education benefits all students without compromising their privacy or security.

Furthermore, the efficacy of AI in personalization heavily depends on the quality of input data. Biased or incomplete data can lead to skewed outcomes, thus reinforcing existing disparities rather than dismantling them. Rigorous standards and audits are necessary to ensure that AI systems act fairly and equitably.

Technological accessibility is another hurdle. Schools and institutions must invest in robust infrastructure to support AI applications. This means not only hardware and software but also training teachers to effectively integrate AI tools into their teaching practices. In this sense, personalized learning isn't a magic bullet but rather a component of a broader educational infrastructure upgrade.

As personalization becomes an integral part of educational paradigms, it's crucial that stakeholders, including policymakers and educational institutions, work collaboratively to define the expected outcomes and ethical boundaries of AI's role in education. Balancing innovation with ethical responsibility is key to ensuring that AI's promise is realized equitably.

Ultimately, the future of personalized learning experiences, bolstered by AI, is bright. Its potential to render education more equitable, engaging, and effective cannot be overstated. As we continue to refine these technologies and practices, the dream of a world where learning is tailored to the learner—allowing everyone to shine in their own light—moves ever closer to reality.

Intelligent Tutoring Systems

As education evolves through the transformative influence of artificial intelligence, intelligent tutoring systems (ITS) stand out as one of the most promising innovations. These systems harness the power of AI to mimic personalized teaching experiences, offering an individualized learning experience tailored to each student's unique needs and pace. It's a revolutionary shift from traditional, one-size-fits-all pedagogies toward a dynamic, student-centered model that could redefine knowledge acquisition.

At the core, intelligent tutoring systems operate by adapting to the learner's needs in real time. Using sophisticated algorithms, these systems assess a student's strengths and weaknesses, curating educational content that aligns with their current level of understanding. This adaptability not only accelerates learning but also enhances retention, leading to a deeper, more sustainable comprehension of the material. The ability to provide immediate feedback and instant clarification on misconceptions fosters a supportive learning environment where students can thrive.

We've all experienced the frustration of trying to grasp complex concepts in a crowded classroom, where individual attention is scarce. Intelligent tutoring systems address this by offering a personalized tutor to every student—no matter where they are. This individualization is a game-changer for inclusivity, allowing learners from diverse backgrounds and varying educational levels to engage meaningfully with their studies.

The design of intelligent tutoring systems is deeply rooted in cognitive science. By understanding how the human brain processes and retains information, these systems are designed to optimize instructional methods and content delivery. This integration of AI and cognitive psychology models allows ITS to not only present information but also engage students in active learning processes, such as problem-solving and critical thinking.

The potential of ITS doesn't stop at academic subjects. These systems can also incorporate soft skills training, such as collaboration, communication, and problem-solving—skills that are increasingly valued in today's global workforce. By simulating scenarios that require these skills, intelligent tutoring systems provide experiential learning experiences that traditional settings often overlook.

Furthermore, intelligent tutoring systems have the capability to address educational disparities. In regions with limited access to high-quality educational resources, ITS offer a viable, scalable solution. With internet connectivity, underserved communities can access world-class education, fostering equity and empowerment. These systems can be multilingual and culturally adaptive, ensuring a wide reach across the globe.

On the technology front, the development of ITS employs machine learning, natural language processing, and even elements of gamification to enhance user engagement. Machine learning algorithms continually refine their assessments and recommendations based on user interactions, while natural language processing allows for more intuitive communication between student and system. Gamification components turn learning into an interactive experience, making education not just effective but also enjoyable.

The implementation of intelligent tutoring systems also sparks a wider conversation regarding the role of educators in the AI-enhanced classroom. Teachers become facilitators, guiding students and evaluating their interactions with AI-based systems. This synergy between human insight and AI capability can enhance the educational experience, allowing teachers to focus on fostering creativity, critical thinking, and emotional intelligence—areas where human educators excel.

Nonetheless, realizing the full potential of ITS requires overcoming certain challenges. Privacy concerns, data security, and the ethical use of AI in education are critical issues that need addressing. It's essential to ensure student data remains secure and that AI behaviors are transparent, unbiased, and aligned with educational goals.

The future of intelligent tutoring systems is rich with potential. As technology advances and becomes more sophisticated, these systems will increasingly contribute to educational innovation. With continued research and development, intelligent tutoring systems can revolutionize how we learn, making education more accessible, effective, and enjoyable than ever before.

In this regard, educators, technologists, and policymakers must collaborate to integrate these systems harmoniously into educational ecosystems. By doing so, we can ensure a future where learning is enhanced by intelligent technology, and each student has the opportunity to realize their full potential.

Chapter 7:
AI Agents in Transportation

The integration of AI agents into the transportation sector marks a pivotal shift toward a future where mobility is more efficient, safer, and increasingly autonomous. By orchestrating advanced algorithms, these agents are reshaping how we navigate our world. They drive innovations in autonomous vehicles, challenging the existing paradigms of car ownership and road safety by reducing accidents with their precise, real-time decision-making capabilities. Beyond this, AI agents redefine traffic management by fine-tuning traffic flow and reducing congestion, thereby cutting down on emissions and travel time. As we step into the era of smart public transport, AI optimizes routes and schedules, promising a seamless travel experience. AI's role in transportation is not just about moving people and goods more effectively; it's about harmonizing human life and technology, paving the way for sustainable urban living, and notably shifting our transportation narrative from human-driven to AI-accompanied journeys. The transformative potential of AI in transportation signals a future where achieving balance between technological prowess and urban needs is no longer a distant reality but a present-day undertaking.

Autonomous Vehicles and Traffic Management

In the ever-evolving landscape of AI, autonomous vehicles stand as a beacon of innovation, promising to redefine the way we approach transportation and traffic management. These self-driving marvels, powered by an array of sensors, algorithms, and machine learning capabilities, aim to enhance road safety, reduce congestion, and minimize environmental impacts. Though the journey towards fully autonomous vehicles is complex and fraught with challenges, the potential benefits offer a compelling vision for the future.

At the heart of autonomous vehicle technology lies the intricate interplay of sensors and machine learning. Lidar, radar, and cameras collect data from the vehicle's surroundings, which is then processed by advanced AI algorithms to make real-time driving decisions. These systems must not only detect and interpret traffic signals, lanes, and potential obstacles but also predict the behavior of other road users. It's an astounding achievement, yet one that demands constant refinement to ensure safety and reliability.

The influence of autonomous vehicles extends far beyond the vehicles themselves. They are poised to transform entire infrastructure systems, requiring enhancements in digital traffic management and urban planning. Technologies like connected vehicle environments, where vehicles communicate over high-speed networks, pave the way for an intelligent transportation infrastructure that can dynamically manage traffic flow and reduce bottlenecks. This synergy between autonomous vehicles and smart infrastructure holds promise for streamlined traffic management, decreased travel times, and a reduction in the frequency and severity of accidents.

The economic implications are equally profound. Autonomous vehicles could reshape industries dependent on human drivers, leading to both disruptive and constructive upheavals. The logistics and delivery industries are anticipated to undergo significant transformations as companies strive for efficiency with self-driving trucks and drones. Moreover, there's the potential for job reformation, where the human workforce transitions from driving roles to more technical or supervisory positions in vehicle management and maintenance.

Public sentiment and regulatory environments are crucial factors in the deployment of autonomous vehicles. Trust in the safety and efficacy of these vehicles is imperative; thus, rigorous testing and transparent communication will be key. Legislators around the globe are tasked with creating adaptive policies that ensure both innovation and public safety. In regions where regulatory frameworks are evolving favorably, we can expect to see more widespread adoption of autonomous vehicles.

Autonomous vehicles could solve persistent urban challenges. Cities plagued by congestion might find relief through improved traffic systems and reduced vehicle ownership rates. Additionally, autonomous ride-sharing services can lead to more efficient vehicle use, trimming down the number of cars on the streets and helping cities reclaim space from parking lots and garages. Furthermore, the environmental impact can be mitigated through optimized routes and more consistent driving patterns, which would reduce emissions and energy consumption.

However, the road to autonomous driving supremacy is riddled with technical and ethical challenges. Ethical driving decisions, such as those concerning collision avoidance, are intensely complex and require multifaceted considerations. Researchers must grapple with the moral dilemmas of choosing the lesser evil in scenarios like unavoidable accidents. Ensuring that autonomous systems learn to prioritize human life requires collaborative efforts between technologists, ethicists, and policymakers.

Data security and privacy are paramount in vehicles that operate with vast amounts of personal and environmental data. Protecting these sensitive data streams from cyber threats is essential to maintaining consumer confidence and securing the broader transportation network. The development of robust cybersecurity protocols is just as critical as the technological advancements in AI driving systems themselves.

As historical barriers to adoption—such as technology limitations, high costs, and societal skepticism—begin to crumble, autonomous vehicles promise to become a mainstay of modern life. Their emergence brings forth opportunities for innovation in adjacent fields, such as AI-driven traffic simulation models, which help urban planners and policymakers anticipate the impacts of self-driving cars on existing cityscapes.

Education and training will also play pivotal roles in preparing societies for this transformation. As AI and autonomy become more integrated into everyday life, developing a curriculum that prepares individuals to work alongside and with these technologies is vital to the transition's success.

The horizon for autonomous vehicles and traffic management is one brimming with potential. While challenges remain, the march towards a future where transportation is safer, more efficient, and environmentally friendly continues unabated. As we collectively strive to unlock these technological possibilities, the vision of a seamlessly integrated autonomous mobility ecosystem moves closer to becoming a reality.

AI and the Future of Public Transport

As we stand on the cusp of a transportation revolution, AI is poised to redefine how we move through our cities and beyond. Public transport systems, which form the backbone of urban mobility, are beginning to integrate AI technologies that promise to enhance efficiency, convenience, and sustainability. This transformation is not just about more buses or trains but about creating a smart, interconnected network that meets the evolving needs of the public.

AI's role in public transport is multifaceted. One significant impact is its ability to optimize route planning and scheduling. By analyzing vast datasets from traffic patterns, weather conditions, and passenger demand, AI systems can create dynamic routes that save time and resources. This means fewer delays, less congestion, and a smoother experience for commuters.

Furthermore, AI technologies are enhancing the safety and reliability of public transit. Predictive maintenance, for example, leverages machine learning algorithms to forecast equipment failures before they occur. This insight allows operators to perform timely maintenance, reducing the risk of breakdowns and ensuring a more reliable service. These proactive measures not only prevent disruptions but also significantly cut down on repair costs.

Innovation in AI is also dramatically improving passenger experiences. From real-time updates on transit schedules to personalized journey recommendations, AI-driven applications are transforming how individuals interact with public transportation. Voice-activated digital assistants can offer ease for those unfamiliar with a transit system, helping them find the best routes with minimal hassle.

Accessibility remains a crucial concern, and AI is making strides here as well. Systems equipped with natural language processing (NLP) can provide crucial information in multiple languages, catering to diverse urban populations. Additionally, AI-based applications can assist visually or hearing-impaired passengers by offering navigation support and alerts specific to their needs, thereby making public transport more inclusive.

Environmental impact is a pressing issue that AI can help address by promoting smarter, more sustainable transit solutions. AI has the potential to decrease pollution levels by optimizing vehicle speeds and managing energy consumption. Electric and autonomous buses managed by AI can further reduce emissions, contributing to cleaner urban environments.

Moreover, AI is indispensable in the growing field of autonomous public transport. While fully autonomous buses might seem futuristic, pilot projects already demonstrate their viability. These AI-controlled vehicles can adhere strictly to traffic regulations and respond rapidly to changing road conditions, potentially enhancing travel safety while reducing human error.

Another exciting prospect is the integration of AI with public transport for improved urban planning. Data-driven insights from AI systems can inform the design and expansion of transport infrastructures to meet current and future demands efficiently. This dynamic approach helps cities accommodate growing populations without being daunted by overwhelming traffic challenges.

As public transport systems continue to evolve with AI integration, there are considerable challenges to tackle. Privacy concerns loom large, as the massive amount of data collected for AI operations must be handled responsibly. Balancing progress with ethical considerations remains paramount to ensure that technological advancements in public transport benefit society as a whole.

There's also the question of investment and infrastructure. Shifting from traditional to AI-enhanced public transport systems requires considerable funding and policy support. Governments and private stakeholders need to collaborate closely to ensure these technologies are deployed equitably and effectively.

Looking ahead, AI will undoubtedly play a crucial role in coping with the unforeseen challenges of urban mobility. Whether through decreasing traffic congestion, improving commuter safety, or minimizing environmental impacts, AI holds the promise of crafting a more intelligent and responsive transportation ecosystem. Public transport systems will no longer be static networks but dynamic entities that evolve alongside the communities they serve.

In conclusion, the future of public transport influenced by AI is not just a possibility but an emerging reality. It invites us to rethink how we design our cities and move within them. As we advance, the challenge will lie in embracing these technologies while tackling the ethical and infrastructural complexities they present. Only by navigating these hurdles can AI's transformative potential in public transport fully materialize, offering a blueprint for smart urban mobility in the 21st century and beyond.

Chapter 8:
AI Agents in Finance

AI agents are at the forefront of transforming the financial sector, bringing a blend of speed, accuracy, and adaptability that human operators struggle to match. Central to this revolution are robotic trading systems, which leverage vast datasets and sophisticated algorithms to execute trades at lightning speed. These systems often use techniques like pattern recognition and predictive analytics, enabling them to anticipate market movements and make informed trading decisions with remarkable precision. Furthermore, AI-driven fraud detection mechanisms are redefining security in finance by monitoring and analyzing transactions in real-time, swiftly identifying anomalies that could signify fraudulent activities. The result is a financial environment that is not only more efficient but also more resilient against threats, capable of adapting to ever-changing market conditions with unprecedented agility. As AI agents continue to advance, their capacity to optimize financial processes and enhance decision-making elevates them as pivotal players in the future of global finance.

Robotic Trading Systems

In the high-stakes world of finance, robotic trading systems are redefining the dynamics of stock markets. Combining the speed and precision of AI with the capacity to analyze vast datasets, these systems execute trades far faster than any human could. It's this speed that marks the core appeal of algorithmic trading. Gone are the days when traders relied solely on intuition and manual analysis; now, sophisticated algorithms scrutinize market data in real-time, offering recommendations or even executing trades autonomously.

These systems leverage a range of advanced techniques. From machine learning models that predict price movements based on historical trends to deep learning strategies that comprehend unstructured data like news reports, robotic trading systems incorporate diverse AI capabilities. Some utilize sentiment analysis to gauge market sentiment from social media and news, while others implement reinforcement learning to adapt trading strategies dynamically. The common thread? They're geared for optimal accuracy and efficiency.

Importantly, robotic trading systems extend beyond simple buying and selling. They play a vital role in risk management, offering unprecedented insights into potential market volatilities. By continuously analyzing market conditions, these systems can identify potential risks, flagging them for human oversight or automatically adjusting portfolios to minimize exposure. This foresight allows institutional investors to safeguard their assets while maximizing returns, a compelling proposition that has increased the demand for AI-driven trading solutions.

While the benefits are clear, robotic trading systems are not without challenges. One significant concern is the 'black box' nature of many AI models, where the decision-making processes are opaque. Investors and regulators alike are cautious about how these decisions are made and what's influencing them. There's a pressing need for transparency in algorithmic trading, prompting ongoing research into explainable AI techniques. These efforts aim to make the inner workings of AI models more accessible, ensuring that stakeholders can trust and understand automated decisions.

The rapid development of robotic trading systems raises questions about market stability. High-frequency trading, driven by these algorithms, can exacerbate market volatility, particularly in times of economic uncertainty. Sudden spikes or declines can trigger automatic sell-offs, leading to a self-reinforcing cycle of volatility. To mitigate these risks, regulators are exploring safeguards, such as circuit breakers, to temper these systems' rapid actions and promote market equilibrium.

Regulatory measures play a crucial role in setting ethical guidelines and ensuring fair practices. In an arena where milliseconds can decide millions, the potential for market manipulation by sophisticated algorithms is a genuine concern. Regulative bodies seek to foster a balance between allowing innovation and protecting market integrity. Striking this balance requires a multi-faceted approach—encouraging transparency, implementing stringent compliance standards, and fostering international cooperation to address the global nature of financial markets.

Despite these challenges, the financial ecosystem continues to embrace robotic trading systems. Their impact is not limited to elite trading firms; retail investors increasingly gain access to algorithmic tools, democratizing investment strategies that were once the realm of Wall Street giants. Robo-advisors use similar technologies to provide tailored investment advice, bridging the gap between complex algorithms and everyday investors. This democratization fosters inclusivity, enabling more individuals to participate in wealth creation with informed decision-making.

The future promises further innovations. Quantum computing, for instance, could transform trading algorithms by solving complex problems at unprecedented speeds. Augmenting AI with quantum technologies holds the potential to revolutionize predictive analytics, offering deeper insights and opening new horizons for market participants. This synergy is at the frontier of financial technology, hinting at possibilities that could reshape trading paradigms.

Robotic trading systems are a testament to AI's transformative power in finance, illustrating both its potential and complications. The road ahead involves unpacking sophisticated algorithms to ensure transparency, crafting regulations that protect markets, and integrating innovations thoughtfully to foster sustainable growth. As these systems evolve, their role in shaping the finance sector—and the global economy—demands careful stewardship.

Fraud Detection Mechanisms

In the ever-evolving landscape of finance, fraud detection stands as a linchpin for maintaining integrity and trust. AI agents have revolutionized this domain with unparalleled precision and speed. Traditional methods of detecting fraudulent activities, while foundational, often relied on linear processes that could be bypassed by savvy fraudsters. However, AI agents, with their ability to learn and adapt, offer a more robust defense system.

Modern AI-driven fraud detection mechanisms leverage machine learning algorithms to analyze vast amounts of transactional data. These algorithms identify patterns that may hint at deception— patterns that a human eye might easily overlook. Such systems are adept at comparing new data against historical data to spot anomalies. For instance, if a customer's spending habits suddenly change or if a transaction originates from an unusual geographic location, the AI system flags it for further review.

Neural networks are at the core of many AI fraud detection systems. They function by mimicking the human brain's structure to process information. These networks are particularly useful for their capability to detect subtle patterns and correlations across extensive datasets. An essential feature of them is their ability to improve continuously; as more data becomes available, these systems refine their criteria for what constitutes suspicious activity.

Consider the fraud detection system of a global bank. Such a system processes millions of transactions daily, searching for deviations from established norms. AI agents ensure these systems react to potential threats in real time, reducing the response time from hours to mere seconds. This ability is crucial when dealing with large-scale fraud schemes, where speed is of the essence.

Moreover, AI agents are exceptionally proficient in reducing false positives—erroneously flagged transactions that are legitimate. Traditional systems could inundate human analysts with alerts that required verification, most of which were not instances of fraud. Through learning patterns of verified transactions, AI agents enhance accuracy, ensuring that more focus is placed on genuine threats. This efficiency saves businesses substantial resources and improves customer satisfaction by minimizing unnecessary disruptions.

Another dynamic feature of AI in fraud detection is its adaptability. As fraudsters develop new techniques, AI agents evolve concurrently, updating their parameters to include the latest fraudulent indicators. This is a stark contrast to earlier static systems, where rules had to be manually adjusted, often leaving a window of vulnerability open during the update process.

AI agents also play an instrumental role in enhancing customer verification processes. Traditional systems often relied on simple passwords or security questions, which are increasingly inadequate. AI enhancements facilitate multi-factor authentication processes and biometric verification. Thus, they strengthen security without compromising user experience, a critical balance in today's digital age where convenience is paramount.

Furthermore, collaborative intelligence between AI and human analysts enriches fraud detection mechanisms. While AI excels at pattern recognition and processing large-scale datasets, human intuition remains invaluable, particularly in nuanced cases where context matters. A synergy of AI's computational prowess with human insight produces a sophisticated fraud detection framework capable of addressing complex fraud scenarios.

Natural language processing (NLP) complements these mechanisms by analyzing communications for signs of fraud. This technology assesses text from emails, messages, and even call transcripts. By recognizing specific phrases or emotional tones that correlate with deceitful intent, NLP-augmented AI systems provide a broader shield against deception, particularly in schemes where social engineering is involved.

Despite their sophistication, AI-based fraud detection mechanisms come with their own set of challenges. Bias in data or AI training processes can lead to discriminatory practices, inadvertently targeting specific groups unfairly. Therefore, it is essential to incorporate fairness and transparency into AI system design. Regular audits and the inclusion of diverse data sets during the training phase help mitigate such risks.

The future of fraud detection in finance lies in fortifying these AI systems with more granular data inputs and expanding their interconnectivity across platforms. Real-time data sharing between financial institutions globally could lead to an even more robust deterrence system. Blockchain technology, combined with AI's analytical power, could provide an indelible and transparent record of transactions, further deterring fraud attempts.

In conclusion, AI agents play a transformational role in advancing fraud detection mechanisms in finance. Their ability to detect intricate patterns, adapt to new threats, and collaborate effectively with human expertise makes them indispensable allies. As technology continues to evolve, the integration of AI agents in fraud detection will only deepen, offering the financial world renewed strength against ever-present fraud risks.

Chapter 9:
Ethical Challenges of AI Agents

As we delve into the realm of intelligent automation, the ethical challenges posed by AI agents demand our attention with increasing urgency. These challenges are not mere hurdles; they're pivotal considerations that shape the moral fabric of our technological advancement. Balancing innovation with ethical responsibility is no small feat. Privacy concerns arise as AI agents burrow deeper into personal data, making the protection of individual rights paramount. Furthermore, the bias entrenched in AI systems reflects societal prejudices, raising questions about fairness and equality. The responsibility lies with developers, policymakers, and society to ensure transparency, accountability, and ethical vigilance in AI implementation. Only through this concerted effort can we foster an AI-driven future that aligns with humanity's core values while harnessing the incredible potential these agents offer.

Privacy Concerns with Intelligent Automation

As the capabilities of AI agents continue to evolve, so do the intricacies of their integration into our daily lives and industries. A principal challenge within this landscape is the issue of privacy, especially as intelligent automation becomes more pervasive. How these advanced systems interact with personal data and utilize it to deliver seamless and personalized experiences raises significant privacy concerns.

AI agents process vast amounts of data to function effectively. This data, often derived from personal information, is crucial for these systems to understand and predict user needs. For instance, digital assistants rely on speech recognition and natural language processing to respond accurately to user queries, necessitating access to personal interactions and preferences. While this capability enhances user experiences, it also begs the question: who controls the data, and how securely is it protected?

Imagine walking into your smart home, and your virtual assistant anticipates your every need, from coffee preferences to playing your favorite music. This seamless integration is powered by algorithms that learn from your behaviors and choices. However, the aggregation and analysis of such intimate data points may lead to unintended privacy intrusions, especially if there's a lack of transparency in data collection practices.

One of the most pressing concerns is data ownership. In many cases, users are unaware of how their data is being used or shared. This lack of clarity is not only unsettling but poses potential risks if data ends up in the wrong hands. The vulnerability of data breaches looms large as attackers continually develop new strategies to exploit weaknesses in security systems.

Moreover, the deployment of AI in various sectors, such as healthcare and finance, necessitates stringent privacy measures. In healthcare, AI-driven diagnostic tools require access to sensitive patient information, which, if mishandled, could lead to violations of patient confidentiality. Similarly, in finance, AI agents assessing credit scores and fraud detection systems analyze monetary transactions and personal data, demanding robust mechanisms to ensure data protection.

Addressing these concerns necessitates a multidimensional approach. Developers and vendors must adopt privacy-by-design principles, embedding data protection into the lifecycle of AI systems. This involves not only designing systems with user privacy in mind but also fostering transparency about how data is collected, processed, and utilized. Users should have clarity and control over their data, with options to manage permissions and data sharing preferences easily.

Informing and educating users about privacy policies and data usage is another critical aspect. Often, users click through lengthy terms of service agreements without fully understanding the implications. Simplified explanations and easy-to-navigate controls can empower users, allowing them to make informed decisions about their data.

Additionally, regulatory frameworks play a pivotal role in safeguarding privacy. Governments and regulatory bodies worldwide are recognizing the importance of instituting standardized policies for data protection in AI applications. Frameworks such as the General Data Protection Regulation (GDPR) in Europe have set precedents, requiring companies to be more accountable in their data handling practices. Similar policies are being considered and implemented in other regions to create a global standard for privacy in intelligent automation.

Yet, regulations alone may not suffice. As AI agents become integral to societal functions, the discussion on privacy must evolve alongside technology advancements. Stakeholders, including tech companies, policy-makers, and consumers, need to engage in ongoing dialogues to navigate emerging privacy challenges. Such collaborations can lead to the development of innovative solutions that balance technological progress with ethical considerations.

An essential aspect of this dialogue is recognizing the potential for AI systems to inadvertently perpetuate surveillance. As AI agents are deployed in public spaces—think smart cities equipped with surveillance cameras and sensors—concerns about pervasive monitoring and the erosion of anonymity arise. Striking a balance between security needs and individual privacy rights becomes paramount in ensuring that intelligent automation serves societal needs without infringing on civil liberties.

In light of these concerns, the development of privacy-preserving technologies is gaining momentum. Techniques such as differential privacy, federated learning, and encryption are being integrated into AI systems to minimize data exposure and protect user identity. These technological advancements signify a critical step towards reconciling the utility of intelligent automation with the imperative of privacy preservation.

Looking ahead, the conversation around privacy concerns with intelligent automation will likely expand, becoming more sophisticated as AI systems become more entrenched in our lives. As these systems gain cognitive abilities akin to human reasoning and decision-making, the stakes of privacy risks will undoubtedly increase. Proactively addressing these challenges will be crucial to maintaining public trust in AI technologies and ensuring their ethical and responsible use.

Ultimately, the potential of intelligent automation is boundless, offering prospects that can significantly enhance our quality of life. However, realizing this potential hinges on our ability to manage and mitigate privacy concerns effectively. By fostering a culture of accountability and standardized privacy practices, we not only protect individual rights but also pave the way for AI systems to thrive as trusted agents of innovation and progress.

Addressing Bias in AI Systems

In a world increasingly influenced by artificial intelligence, the ability of AI systems to act impartially is under intense scrutiny. Bias in AI systems is not just a technical flaw; it's an ethical challenge that shapes the fairness and equity of the decisions made by these agents.

Understanding and addressing this bias is crucial for creating systems that can be trusted across diverse user bases and applications.

Bias can creep into AI systems at various stages, from the data used for training to algorithms' design and deployment environments. It's essential to acknowledge that AI systems reflect the values and prejudices present in their training data. If historical data is tainted with bias based on race, gender, or socioeconomic status, the AI system will likely perpetuate those biases in its decision-making process. This can lead to exclusionary practices or unfair treatment of certain groups.

Moreover, the complexity of AI models can obscure the underlying reasons for their decisions, making it difficult to identify and correct bias. Known as the "black box" problem, this opacity challenges stakeholders to dissect how AI arrives at its conclusions. Without transparency, ensuring that AI systems operate without bias remains a significant hurdle.

Yet, tackling bias isn't only about technical solutions; it requires a societal approach that involves policy-making, diverse participation in AI development, and adherence to ethical guidelines. Agencies, developers, and companies need to recognize their role in reshaping AI to be more equitable. Diverse teams, inclusive of various races, genders, and backgrounds, are pivotal in challenging existing assumptions within AI systems and proposing balanced solutions.

The role of regulation is paramount in setting standards that demand fairness and transparency. Regulatory efforts should aim to create guidelines that govern the development and deployment of AI, ensuring systems are tested for bias before they are released into the public domain. These regulations need to be adaptable to incorporate the rapid technological advancements within AI while remaining steadfast in their ethical purposes.

In practice, bias mitigation could involve techniques like reweighting data points to ensure that underrepresented groups are adequately considered. Additionally, fairness constraints can be integrated into algorithms to ensure that the outputs meet specific equitable criteria. These strategies must be woven into the fabric of AI development from the onset, rather than as an afterthought.

Transparency and accountability are another critical part of addressing bias. Developers and organizations must be tasked with documenting the processes and decisions behind AI system designs. Such documentation should be accessible not only to technical experts but also to the general public, ensuring widespread understanding and trust in AI systems.

One promising avenue is the development of explainable AI (XAI) systems, which aim to shed light on how decisions are made. By providing insights into the decision-making processes of AI, XAI allows users to detect and address bias more effectively. Moreover, there is a growing movement advocating for AI audit trails, which would document the decision paths of AI systems, offering a retrospective examination of actions to identify and rectify biases.

The integration of cultural competence into AI systems presents another frontier in overcoming bias. Systems need to be sensitive and adaptable to cultural differences, reflecting the nuanced ways people from various backgrounds interact with technology. By acknowledging and incorporating different cultural perspectives, AI systems can become more inclusive and reduce bias.

Although considerable challenges remain, steps are being taken globally to address AI bias. Initiatives and research projects dedicated to exploring bias in AI are budding across the world, fostering collaborations between academia, industry, and policy-makers. This collective effort holds the potential to facilitate the development of more nuanced, ethically sound AI systems.

The journey to addressing bias in AI systems is ongoing, demanding vigilance, commitment, and innovation. The advancement of AI technology should parallel an equally vigorous pursuit of ethical responsibility. As AI becomes more embedded in the fabric of society, the repercussions of biased decisions could become even more pronounced. Thus, creating unbiased, equitable AI is not just about enhancing technical performance; it's about safeguarding human values and ensuring a future where technology serves all equitably.

The essential question we face is not whether we can eliminate all biases but how we can create AI systems that actively recognize and counteract existing disparities. This mission compels us to reflect on the kind of technological future we desire and work collectively towards achieving it, ensuring that AI evolves as a force for good, cognizant of the diversity and complexity of human life.

In conclusion, addressing bias in AI systems intersects with technical innovation and social necessity. It challenges us to expand our horizons of understanding, accepting that AI development must be intrinsically linked with ethical considerations. As we strive to develop intelligent agents that embody our highest aspirations for a just society, we stand on the precipice of shaping the future of AI in a manner that is not only intelligent but profoundly humane.

Chapter 10:
Regulatory Landscape of AI Agents

The regulatory landscape of AI agents is as dynamic and intricate as the technologies themselves, necessitating a coherent synthesis of innovation and legislation. With AI's rapid advancements, policymakers face the arduous task of crafting regulations that both encourage technological growth and safeguard societal interests. The challenge lies in striking a balance where innovation thrives without compromising ethical standards, privacy, and security. Initiatives at the global level aim to establish comprehensive standards, fostering international collaboration to address the multifaceted implications of AI. These efforts include aligning AI policies with human rights, ensuring transparency in AI decision-making processes, and actively engaging diverse stakeholders in shaping the future of AI governance. As AI agents increasingly permeate various sectors—from finance to healthcare—the development of robust regulatory frameworks becomes paramount, not only to mitigate potential risks but also to unlock the full potential of AI in driving societal progress. In essence, navigating this regulatory landscape requires a forward-thinking approach that is both adaptive and inclusive, paving the way for an ethical and responsible AI-driven future.

Navigating AI Policies and Regulations

The rapid evolution of artificial intelligence (AI) has heralded a transformative era across various sectors. Yet, this incredible progress also spotlights a complex web of policies and regulations necessary to govern these revolutionary technologies. As AI agents integrate deeper into our lives, governments and institutions worldwide are grappling with crafting regulations that both foster innovation and protect societal interests. Adequate regulation is fundamental in ensuring ethical development, deployment, and utilization of AI agents.

One of the critical challenges in regulating AI agents is their diverse applications. AI is not a monolithic technology but rather a constellation of systems that operate across industries—ranging from healthcare to finance, and transportation to entertainment. This broad applicability means that a one-size-fits-all regulatory approach is unfeasible. Instead, policymakers must tailor regulations to specific sectors while ensuring a cohesive framework that addresses overarching ethical concerns such as bias, privacy, and accountability.

Developing effective AI policies requires an understanding of the underlying technologies that power these agents. Machine learning, natural language processing, and neural networks are just a few core components that need regulation. Each of these technologies presents unique challenges. For instance, machine learning models often function as "black boxes," making it difficult to interpret decisions and outcomes. Crafting legislation that demands transparency in AI systems while still allowing room for innovation is a balancing act that regulators must master.

Moreover, privacy concerns are paramount when developing AI policies. As AI agents become more adept at gathering and analyzing data, ensuring individual privacy becomes a complex task. Regulations need to define boundaries for data collection and usage, compelling organizations to adopt stringent data protection measures. Legislation such as the General Data Protection Regulation (GDPR) in Europe has set a precedent for data privacy, emphasizing user consent and data security in the context of AI applications.

In addition to privacy, addressing bias in AI systems is a pivotal aspect of regulation. AI agents are only as unbiased as the data they are fed, making them susceptible to reflecting and amplifying existing societal inequalities. Policymakers must enforce regulations that mandate diverse and representative data sets in training AI models, as well as measures to continually audit and rectify unfair biases. This ensures that AI agents make equitable decisions across different domains.

Accountability is another cornerstone in AI regulation. As AI systems assume roles that carry significant decision-making power—such as in autonomous vehicles or healthcare diagnostics—determining liability in cases of failure or harm becomes crucial. Regulations must delineate the responsibilities of AI developers, users, and manufacturers to ensure accountability and provide clear recourse for affected parties. This is no trivial task, as pinpointing accountability in complex, self-learning systems often resembles untangling a Gordian knot.

The global nature of AI technology further complicates regulatory efforts. Differing political, cultural, and economic contexts across nations lead to diverse regulatory landscapes. Even though some countries take the lead with progressive AI policies, global coordination remains essential to prevent regulatory fragmentation. International organizations and consortiums play a critical role in harmonizing standards, fostering cross-border collaborations, and establishing universal guidelines for AI ethics and governance.

One of the significant hurdles in creating global AI policies is the pace at which technology outstrips regulation. AI development is swift and often outpaces the legislative process, resulting in a "regulation gap." Policymakers need to adopt agile regulatory models that can adapt to technological advances without stymieing innovation. Sandboxing, for instance, provides a controlled environment for developing and testing AI solutions under temporary regulatory dispensations, encouraging innovation while maintaining oversight.

Public and private partnerships are invaluable in navigating AI policies and regulations. Technological stakeholders often have specialized knowledge that can aid governmental bodies in crafting effective governance frameworks. Collaborations between tech companies, academia, and regulatory agencies foster an environment of shared expertise and resources, ensuring policies are both informed and practical. Encouraging open dialogue and collaboration can mitigate the risks associated with regulatory uncertainty.

Furthermore, an informed public plays a crucial role in shaping AI policy. Raising awareness about AI's potential and pitfalls empowers citizens to engage in meaningful discourse about the kinds of regulations they support. Public consultations and stakeholder engagements can offer valuable insights into societal needs and ethical considerations, providing a grassroots perspective that enriches the policy-making process.

As we chart a course through this ever-evolving landscape, it's imperative to keep ethical considerations at the forefront of AI policy formulation. While AI agents open a wealth of opportunities, they also raise profound ethical questions about autonomy, human agency, and the evolving dynamics of work and life. Regulatory frameworks must continually evolve to ensure that AI serves humanity positively, respecting fundamental human rights and societal values. Crafting such frameworks is not only a legislative challenge but also a societal imperative that demands collective effort and responsibility.

Global Standards for AI Implementation

As artificial intelligence permeates various sectors worldwide, establishing global standards for AI implementation becomes essential to ensure consistency, safety, and ethical integrity. The rapid evolution of AI technologies has led to a patchwork of initiatives and guidelines across different nations, making global cooperation pivotal. This section delves into the necessity and development of global standards, considering the complex interplay of technology, ethics, and policy.

To begin with, global standards serve as a cohesive framework for harmonizing AI use across borders. With multinational companies deploying AI solutions in diverse regions, discrepancies in local regulations can create challenges. A unified set of standards helps businesses operate smoothly, promoting innovation while maintaining compliance with overarching ethical guidelines. Furthermore, businesses can gain greater trust from consumers when they know that products or services adhere to global standards.

The establishment of these standards, however, isn't a straightforward task. The diverse technological maturity levels and cultural perspectives across countries pose significant hurdles. Nations vary in their approach to privacy, data management, and ethical considerations. For example, what one country considers an invasion of privacy might be deemed acceptable by another. Therefore, these standards must be flexible enough to accommodate cultural differences yet stringent enough to ensure fundamental ethical values remain intact.

To drive the global standardization effort, international organizations and coalitions come into play. Entities such as the International Organization for Standardization (ISO) and the Institute of Electrical and Electronics Engineers (IEEE) are actively working on creating a collaborative platform. Their aim is to bring together stakeholders from government, industry, and academia to converge on robust and adaptable frameworks. Initiatives like the OECD Recommendations on AI represent steps towards international cooperation, emphasizing the importance of transparency, accountability, and fairness in AI systems.

An indispensable aspect of global standards is the emphasis on ethical AI. This involves establishing principles that guide AI toward supporting human rights, promoting fairness, and preventing harm. Ethical considerations extend beyond just programming simple moral guidelines into AI. They require ongoing scrutiny and adaptation as AI systems evolve, ensuring their outcomes align with societal norms and moral acceptability.

One of the corners of global standards is data governance. As AI systems heavily rely on data input, the way data is collected, stored, and used comes under stringent scrutiny. Regulations like the General Data Protection Regulation (GDPR) in the European Union exemplify efforts to safeguard personal data. However, achieving a uniform global data governance framework is challenging due to variances in national legislative priorities. Nonetheless, overarching protocols for data protection can help mitigate risks and enhance public trust in AI technologies.

The role of cross-border collaborations cannot be overlooked in the pursuit of global AI standards. Bilateral and multilateral collaborations facilitate knowledge exchange and foster technological advancements. Countries can learn from each other's successes and failures in AI deployment, paving the way for more comprehensive and resilient standardization frameworks. Collaborative platforms also help address concerns like AI-driven inequality and job displacement, ensuring AI's benefits are distributed equitably across societies.

Moreover, standardization efforts must focus on ensuring the safety and reliability of AI systems. As AI becomes more autonomous, especially in high-stakes applications like healthcare and transportation, comprehensive safety protocols become crucial. Global standards should mandate rigorous testing and validation processes to ascertain AI systems' robustness and mitigate potential risks. Faulty systems could have catastrophic consequences, further highlighting the need for rigorous compliance with established safety norms.

Enforcement mechanisms and compliance checks form the other critical part of global AI standards. Without effective enforcement, even the most well-conceived standards fall short. Regulatory bodies at both international and national levels are vital in ensuring adherence. These bodies need to possess the authority and resources necessary to monitor and enforce AI compliance across different sectors and regions. Regular audits and certification processes can also enhance accountability and transparency in AI implementation.

Finally, the standardization process itself must be adaptive. AI technology is continually evolving, rendering static standards obsolete quickly. Therefore, global standards for AI implementation must be dynamic, with frequent updates informed by the latest technological advancements, ethical considerations, and societal feedback. An adaptive standardization model holds promise as it can evolve alongside technology, providing a robust framework for guiding AI use responsibly and ethically over time.

In conclusion, aligning global standards for AI implementation is a monumental yet indispensable task. It demands collaboration among international bodies, governments, industry leaders, and scientists to establish a cohesive framework that prioritizes ethical integrity, safety, and equitable innovations. Though challenging, these efforts are crucial for shaping an AI-driven future that aligns with global welfare and societal values.

Chapter 11:
The Role of AI Agents in Security

In an era where digital landscapes evolve at a pace previously unimaginable, AI agents are stepping into pivotal roles to fortify security frameworks across all sectors. Their ability to process vast amounts of data rapidly and with an accuracy far surpassing human capacity allows for proactive measures in identifying and neutralizing threats before they manifest. AI agents are not just watchdogs; they're intelligent strategists capable of adapting to sophisticated cyber threats, analyzing patterns, and responding with speed that's reshaping the cybersecurity battleground. While these advancements bolster defenses, they also bring to light complex challenges, especially in balancing surveillance capabilities with privacy concerns. Society stands at a crossroads, where harnessing the power of AI in security necessitates thoughtful integration and careful ethical considerations to ensure that its far-reaching impact enhances safety without compromising individual freedoms.

Cybersecurity Enhancements through AI

Cybersecurity has long been a cat-and-mouse game, where defenders and attackers continually vie for the upper hand. The advent of AI has brought a seismic shift to this dynamic, offering new tools and techniques that both enhance and complicate the landscape of digital security. With cyber threats evolving at an unprecedented pace, leveraging AI capabilities presents a groundbreaking opportunity to build stronger, more resilient security systems that can adapt in real-time.

One of the core strengths of AI in cybersecurity lies in its ability to process and analyze vast amounts of data quickly and efficiently. Traditional methods were often reactive, responding to threats only after breaches occurred. AI, however, can proactively predict and identify potential threats by sifting through enormous datasets to detect patterns indicative of malicious behavior. This shift from reactive to proactive measures represents a game-changer, allowing security systems to anticipate and mitigate threats before they can cause harm.

Machine learning algorithms, a cornerstone of AI technology, play a critical role in this transformation. These algorithms can be trained to recognize the subtle patterns and anomalies that human analysts might miss. They can continuously learn from new data, making them incredibly adept at identifying zero-day vulnerabilities—those previously unknown software flaws that hackers frequently exploit. As a result, organizations employing these technologies can stay one step ahead, patching vulnerabilities before they are widely knew about in the hacker community.

Moreover, AI-driven security systems have demonstrated a capacity for decision-making autonomously, which dramatically enhances incident response times. AI agents can take immediate action to isolate affected systems, cut off compromised connections, or alert security personnel, often resolving issues faster than humanly possible. This speed is crucial in mitigating the extent of damage during an attack, potentially saving companies millions in recovery and restitution costs.

AI is not only a passive defender but also an aggressive predictor of future cyber threats. Predictive analytics allows security teams to forecast potential attacks based on historical data and emerging trends across different sectors. By understanding the nature and source of previous attacks, AI can model potential future scenarios and prepare defensive tactics accordingly. This strategic foresight enables more robust security posture, further complicating the attack landscape for cybercriminals.

The integration of Natural Language Processing (NLP) with AI-based cybersecurity solutions adds another layer of sophistication. NLP can analyze communications over networks, emails, and other channels to detect phishing attempts, which remain one of the most common and damaging types of cyber attacks. By understanding the linguistic patterns of social engineering strategies, AI systems can flag suspicious messages before they even reach the intended recipient, drastically reducing the likelihood of successful phishing attacks.

However, the use of AI in enhancing cybersecurity is not without its challenges. AI systems, especially those incorporating machine learning, require vast amounts of data for training. This necessity raises two significant concerns: data privacy and data integrity. Ensuring that the data used to train these systems is both secure and anonymized is vital to maintain privacy and meet regulatory standards. Moreover, if the datasets are manipulated, it could lead to improperly trained models that might fail to identify threats, or worse, create false positives that could disrupt operations.

There's also the risk that these advanced technologies could fall into the wrong hands. Just as AI can be used to bolster defenses, there's potential for malicious entities to leverage AI's power for nefarious purposes. AI-enhanced cyber attacks, which might use machine learning to better understand and navigate defenses, pose a new kind of threat that security teams must anticipate. This dual-use nature of AI technology emphasizes the importance of ongoing investment in AI research and system hardening.

In the ecosystem of cybersecurity, collaboration between humans and AI is essential. While AI can automate many aspects of threat detection and response, human oversight remains crucial in interpreting results, making judgment calls, and providing the nuanced understandings that machines cannot replicate. The most effective security strategies will weave AI capabilities with human intelligence, creating a hybrid defense system that leverages the strengths of both.

Furthermore, the integration of AI into cybersecurity solutions can lead to more accessible security tools for businesses of all sizes. AI-driven platforms often come with automated features that simplify complex security protocols, making them manageable for companies lacking extensive IT resources. This democratization of advanced security measures helps level the playing field, ensuring that more organizations can protect themselves against increasingly sophisticated cyber threats.

In conclusion, AI's integration into cybersecurity represents a profound enhancement in the ability to protect digital infrastructures worldwide. While challenges and ethical considerations remain, the potential benefits of AI-driven cybersecurity solutions are transformative. By continuing to innovate and evolve these technologies, society can forge a path towards a more secure digital future, where AI acts as a vigilant guardian of information integrity.

Surveillance and Privacy Implications

As the world leans more heavily into the capabilities of AI agents to boost security, there's a pressing need to balance their benefits against potential privacy concerns. AI-driven surveillance systems promise to significantly enhance our security architecture, but they also blur the lines between safety and unfettered intrusion into personal spaces. How, then, do we harness AI's power without sacrificing our fundamental rights to privacy?

While AI agents excel at bolstering cybersecurity measures and responding to threats with impressive speed, they often require access to vast amounts of personal data to function effectively. This necessity sometimes creates an environment where surveillance can cross ethical boundaries. Tools designed for protection might be repurposed, leading to invasive monitoring that could undermine public trust. For this reason, it is crucial that innovations in AI security are matched with equally robust privacy protections.

Advanced AI surveillance systems can process and analyze data far beyond human capabilities. They can scour digital landscapes, identifying potential threats through patterns and anomalies. However, this level of scrutiny can inadvertently scoop up the personal information of those not involved in any wrongdoing. The prospect of being constantly monitored has profound implications for personal freedom. It can stifle free speech and breed a culture of self-censorship when people feel that their every move is being watched.

The rise of facial recognition technology is a perfect example of this dichotomy. While it serves as a powerful tool for law enforcement, aiding in the identification of suspects and speeding up investigations, it also raises substantial privacy issues. Unauthorized access to biometric data could lead to identity theft and other forms of misuse. Furthermore, this type of surveillance can sometimes be carried out without public knowledge or consent, challenging democratic notions of transparency and accountability.

In the quest to create fail-proof security systems, government and tech developers face the challenge of drawing the line between necessary monitoring and unnecessary intrusion. Regulations often lag behind technological advancements, leaving gaps that could be exploited by those with privacy-invasive tendencies. These gaps require urgent attention to ensure that privacy rights are not overshadowed by unwarranted surveillance practices.

Several countries are grappling with these dilemmas, striving to strike a balance in their legal frameworks. In some regions, comprehensive data protection laws have been put in place to safeguard personal information while still allowing for the use of AI in public safety. These laws often mandate that data collection be limited to what is necessary and proportionate to the intended purpose of the surveillance, and that transparency be maintained regarding data usage.

Moreover, strategies such as anonymization and data minimization are being employed to help mitigate privacy concerns. Anonymization ensures that personal identifiers are stripped from datasets, making it more challenging for individuals to be targeted. Meanwhile, data minimization endeavors to restrict the collection of personal information to only what is essential, thereby reducing the risk of potential misuse.

Transparency and accountability in AI-driven surveillance are not just about compliance with the law, however. They are fundamental for building trust with the public. AI systems should incorporate mechanisms that ensure decisions made by these systems are explainable and auditable. Users must be able to understand how and why decisions are made, holding these systems to the same standards we expect from human decision-makers.

Implementing robust cybersecurity measures to protect the information collected by surveillance systems from unauthorized access is just as critical. The occurrence of data breaches can have far-reaching consequences, potentially compromising national security and invading personal privacy on an unprecedented scale. Therefore, safeguarding these systems against external threats must be prioritized.

Despite these challenges, the integration of AI surveillance technologies continues to grow. Businesses, governments, and institutions are eager to exploit their potential to detect and mitigate threats proactively. The economic and social benefits of advanced surveillance are substantial—enabling safer communities, reducing crime rates, and improving service delivery. Nevertheless, as AI agents become more integrated into our security frameworks, society must begin addressing the ethical considerations they introduce.

AI agents have the power to revolutionize security, but their deployment must not come at the cost of privacy and civil liberties. The task before us is not simply to manage the technology itself but to create a culture that prioritizes privacy as a human right. This involves ongoing, dynamic dialogue between technologists, policymakers, and the public to ensure that the right policies are in place to safeguard individual freedoms in an AI-driven world.

In conclusion, the journey of AI agents in the realm of security is just beginning, and so too is the conversation on surveillance and privacy. We find ourselves at a crossroads that demands careful consideration and decisive action. By acknowledging the implications and acting responsibly, we can pave the way for AI-driven security that respects the balance between safety and privacy—ensuring a future where technology serves us, not the other way around.

We must put ethics at the forefront of this evolution, continually evaluating the technologies that hold the promise of a safer world. Only then can we hope to usher in an age where AI agents enhance security without compromising what it means to be free. Privacy and safety are not mutually exclusive, but it requires intention and diligence to achieve both in harmony.

Chapter 12:
AI Agents in Retail

AI agents are radically reshaping the retail landscape, enabling businesses to craft highly personalized customer experiences and optimize operational efficiencies. With advanced algorithms and data analytics, AI-driven systems can deliver tailored marketing strategies that resonate with individual preferences, enhancing customer loyalty and satisfaction. Inventory management, too, sees a transformation, as intelligent agents accurately predict demand patterns and streamline supply chain processes, minimizing waste and ensuring product availability. These dynamic innovations not only drive sales but also foster a seamless shopping journey through intelligent automation. As the retail sector evolves, the integration of AI agents stands as a testament to the power of technology in redefining consumer engagement and operational excellence, envisioning a future where the shopping experience is both intuitive and responsive to ever-changing market trends.

Personalized Marketing Tactics

In the rapidly evolving landscape of retail, artificial intelligence agents stand poised at the forefront of personalized marketing strategies. As digital engagement becomes more nuanced, the ability to craft individualized marketing experiences has shifted from a mere advantage to a fundamental necessity. With the infusion of AI, personalized marketing is no longer about categorizing consumers into broad segments but about tailoring interactions based on intricate behavioral insights and preferences.

AI agents can analyze vast datasets comprising every shopper's interaction within an online ecosystem. This capability transforms consumer data into valuable insights, allowing brands to predict purchasing behaviors with astonishing accuracy. By utilizing machine learning algorithms, AI can discern patterns and trends from historical data, providing retailers with a detailed understanding of each client's likes, dislikes, and purchasing habits. This sophisticated level of insight enables hyper-personalized marketing tactics, creating a customer experience that feels uniquely crafted for each individual.

At the heart of these tactics is the ability to dynamically generate product recommendations. A typical example would be an online retailer using AI agents to suggest products based on a customer's past purchases, browsing history, and even real-time behavior on the website. These recommendations are not static but evolve with each interaction, ensuring that the suggestions remain relevant and appealing. This capability helps retailers maintain consumer engagement by continually offering value and saving customers' time spent searching for desired products.

Furthermore, AI agents play a crucial role in personalized email marketing campaigns, crafting messages that resonate on a personal level. Through natural language processing, AI can tailor the tone and content of email communications to match the recipient's preferences and past interactions with the brand. Whether it's offering discounts on frequently purchased items or announcing a new product line tailored to the consumer's interest, the potential for personalization is limitless. Such targeted communication can lead to increased open rates and higher conversion rates, significantly boosting sales.

Besides direct communications, AI agents also refine customer segmentation. Traditional segmentation relied heavily on demographic information, such as age or location, but AI has revolutionized this approach by integrating psychographic data. By analyzing browsing patterns, social media interactions, and even sentiment analysis from customer feedback, AI can classify customers into highly specific segments based on lifestyle choices, attitudes, and personality traits. This deepened understanding allows marketers to craft messages and offers that truly resonate, fostering a strong brand connection and elevating the customer experience.

Delving further, AI has the potential to enhance in-store experiences through location-based marketing. With the help of AI-driven mobile applications, retailers can send personalized notifications to customers as they walk through a store. Imagine a shopper entering a clothing section and receiving a notification about a personalized discount on their favorite brand. This kind of real-time engagement turns every shopping trip into a uniquely tailored experience, making in-store shopping as personalized as its online counterpart.

The scalability of AI personalization tactics ensures that as businesses grow, their level of customer engagement does not decline in quality. Automation allows these personalized marketing efforts to be maintained across vast customer bases without human intervention, ensuring consistency and efficiency. This scalability is essential for large retailers managing millions of customer interactions daily, allowing them to deliver personalized experiences at every touchpoint.

However, amidst the optimizations, challenges do arise. Privacy concerns are at the forefront, as customers become increasingly wary of how their data is used. Retailers must navigate the fine line between personalization and privacy, ensuring transparency in data usage while delivering valuable experiences. This calls for stringent data protection practices and clear communication with customers about data collection processes, ultimately aiming to enhance trust and loyalty.

Furthermore, the effectiveness of AI-driven personalization may depend largely on data quality. Poor data integration can lead to less accurate insights and misguided marketing efforts. It's crucial for retailers to invest in robust data management systems that ensure accuracy and reliability in the customer data they collect and analyze.

Despite these hurdles, the potential of AI in transforming personalized marketing tactics within retail is immense. Ultimately, these tactics can lead to stronger brand loyalty and higher customer retention rates, driven by the meaningful connections AI helps establish between retailers and consumers. As AI technology continues to advance, the possibilities for personalization will only broaden, setting a new standard for how brands communicate with their audiences.

In conclusion, AI agents infuse personalized marketing with a level of precision and creativity previously unattainable. The profound insight these agents provide enables retailers to cultivate a marketing strategy that is not only efficient but genuinely meaningful. As the retail environment continues to evolve, the strategic implementation of AI-driven personalized marketing will be indispensable for brands aiming to lead in customer engagement and satisfaction, paving the way for innovative, customer-centric retail landscapes.

Inventory Management Solutions

In the vibrant world of retail, effective inventory management can be the difference between thriving and merely surviving. Enter AI agents, the game-changers redefining this landscape. By integrating AI-driven inventory management solutions, retailers can significantly optimize their supply chains, maintain ideal stock levels, and ultimately enhance customer satisfaction.

Inventory inefficiencies, such as overstocking, understocking, and mismanagement, have long plagued retailers, leading to significant financial losses and customer dissatisfaction. Traditional inventory management methods, relying heavily on manual tracking and predictions, often fall short in today's fast-paced market. AI agents, however, bring precision and automation to these processes, transforming how inventory is managed from the ground up.

At the core of AI-driven inventory solutions is the ability to analyze vast amounts of data. By leveraging machine learning algorithms, AI agents can predict demand with remarkable accuracy. These agents analyze historical sales data, current market trends, seasonality, and even the impact of external events to forecast future inventory needs. This predictive capability ensures that retailers have the right products in the right quantities, minimizing both overstocking and stockouts.

Beyond prediction, AI agents bring automation to inventory replenishment. With real-time data analysis, AI solutions can trigger automatic reordering when stock levels dip below predetermined thresholds. This level of automation reduces the need for manual intervention, saving valuable time and reducing human error. In essence, AI agents serve as vigilant custodians of inventory, optimally managing stock levels around the clock.

One of the most compelling aspects of AI inventory management is the capacity for dynamic pricing strategies. By understanding demand patterns and competitor pricing, AI agents can suggest optimal pricing strategies to maximize revenue without sacrificing customer satisfaction. These agents balance competitive pricing with profit margins, ensuring a win-win for both retailers and consumers.

Moreover, AI agents enhance inventory visibility and transparency. With real-time insights into stock levels across different locations, retailers can make informed decisions, such as redistributing inventory to areas with higher demand. This capability is particularly beneficial for retailers with multiple sales channels, as it ensures consistency in inventory levels across both physical stores and online platforms.

Retailers can also harness AI agents to improve the accuracy of order fulfillment. By analyzing logistics data, these agents optimize picking paths and suggest efficient packing strategies, reducing processing times and improving customer satisfaction through quicker deliveries. Enhanced accuracy in fulfillment reduces the likelihood of errors in orders, which can lead to costly returns and negatively impact the customer experience.

Predictive maintenance is another area where AI agents prove invaluable. By analyzing the performance data of inventory management systems, AI solutions can foresee potential equipment failures and suggest maintenance schedules accordingly. This proactive approach minimizes downtime and ensures that systems are always running at optimal efficiency, further streamlining inventory processes.

AI-driven inventory management isn't just about internal efficiencies. It has far-reaching implications for the competitive position of retailers. By optimizing inventory, retailers can respond more agilely to market shifts, capitalize on trends, and improve their overall market presence. In a retail environment where margins are becoming slimmer, these advantages can be critical.

Sustainability is another vital consideration, and AI plays a part in promoting eco-friendly practices in inventory management. By optimizing stock levels and minimizing waste, retailers contribute to reducing their carbon footprint. Efficient inventory systems lead to less overproduction and fewer resources wasted on unsold stock. Such sustainable practices align with growing consumer demand for environmentally responsible business operations.

The implementation of AI agents in inventory management does come with its challenges. These solutions require significant data integration, robust infrastructure, and cross-departmental collaboration to be effective. Initial costs and the need for ongoing maintenance can also be daunting for some retailers. However, the long-term efficiencies, cost savings, and competitive benefits make AI-driven inventory solutions a worthy investment.

In conclusion, the integration of AI agents within retail inventory management offers a transformative approach that elevates efficiency, accuracy, and strategic decision-making. It's a technological evolution that's reshaping the retail landscape, providing a blueprint for how businesses can thrive amidst ever-evolving consumer demands and market dynamics. By embracing AI inventory solutions, retailers are not simply keeping up with the competition; they are setting the stage for a future where intelligent, data-driven decision-making defines success.

Chapter 13:
AI Agents in Entertainment

A I agents have become a transformative force in the entertainment industry, reshaping the way we consume and interact with content. At the heart of this revolution are content recommendation systems, which tailor suggestions to our unique tastes, turning every user experience into a personalized journey. These systems go beyond traditional algorithms, using complex machine learning models that analyze viewing habits, social interactions, and even emotional responses. Meanwhile, AI-driven content creation is ushering in a new era of creativity, where virtual beings can generate scripts, music, and even entire films. These innovative agents are not only enhancing creativity but also democratizing it, allowing anyone to bring their creative visions to life without needing Hollywood budgets. As AI agents continue to evolve, they'll likely redefine what entertainment means, making it more interactive, inclusive, and immersive than ever before. By blending creativity and technology, AI agents are poised to push the boundaries of entertainment, forever altering our engagement with art and media.

Content Recommendation Systems

Have you ever wondered why platforms like Netflix, YouTube, or Spotify seem to know exactly what you'll enjoy next? This near-magical ability to anticipate our preferences is made possible by content recommendation systems. These intelligent systems leverage machine learning and immense data sets to transform hours of user interaction into insightful predictions about user behavior, tastes, and needs. Working in real-time, they sift through vast libraries of content to deliver tailored suggestions, personalizing our digital experiences and often introducing us to options we might never have discovered on our own.

The roots of these systems lie in collaborative filtering, a method that focuses on finding similarities between users or items. The basic principle is simple: if users X and Y have shown similar preferences in the past, X might like something that Y rated highly, but X has not yet encountered. Utilizing algorithms that process these relationships, recommendation systems have perfected the art of linking users to content that aligns closely with their established likes and dislikes. This approach has proven exceptionally effective in platform settings where user bases are large and content inventories expansive.

Content-based filtering represents another prominent technique within recommendation systems. Unlike collaborative filtering, content-based filtering analyzes the properties of items themselves. By understanding what makes a given piece of content unique, these systems can recommend similar items. This method becomes particularly useful when platforms lack extensive user data, harnessing descriptive metadata about content—such as genre, cast, or style—to generate recommendations. Combining content-based and collaborative filtering methods often yields the best results, striking a balance between personalized recommendations and novel discoveries.

Deep learning has significantly enhanced the capabilities of recommendation systems, ushering in a new era of context-aware suggestions. Neural networks, capable of handling complex, non-linear relationships, empower these systems to consider a wider array of data points, from viewing times and device types to historical interests and real-time feedback. This added depth and nuance enable the generation of more sophisticated recommendations that adapt to a user's changing preferences and contexts, offering a dynamic and continuously evolving user experience.

The effectiveness of a recommendation engine often depends on the quality and diversity of the data it receives. Successful platforms invest heavily in collecting and curating comprehensive data, ensuring that recommendations are not only relevant but also diverse, introducing users to fresh perspectives and ideas. The theater may fear homogenization, but when skillfully managed, recommendation systems can be instruments of diversification, emphasizing underrepresented or otherwise overlooked content.

The rise of artificial intelligence in entertainment has fostered unprecedented user engagement, but it also raises important ethical questions. Concerns about privacy and the extent to which personal data is harvested and utilized for these systems frequently surface in public discourse. Users rightfully expect a transparent and secure processing of their information, making ethical data management practices and clear communication essential. Platforms must strive to balance personalization with privacy, ensuring users feel their choices enhance their experience without compromising security.

Despite these challenges, the potential benefits of recommendation systems are considerable, not only from a business perspective but also for user satisfaction and retention. By creating a tailored experience, platforms can drive deeper engagement, leading to longer user sessions and increased content consumption. This symbiotic relationship benefits creators and consumers alike, driving viewership and affording more exposure to a broad range of content creators.

As our digital media consumption grows, so too does the responsibility to ensure recommendation systems foster authentic digital connection and learning. In classrooms, e-learning platforms have adopted similar systems, allowing personalized suggestions that cater to a learner's pace and comprehension. In healthcare, AI-driven recommendations can enhance treatment plans. The broader implications for cross-industry adoption suggest a world increasingly shaped by intelligent personalization.

Looking forward, the innovation horizon for content recommendation systems is vast. With advancements in AI, future systems could leverage real-time environmental and behavioral cues to offer hyper-contextualized recommendations, further blurring the lines between humanity and technology. Imagine a system that not only understands your preferences but also perceives your mood, curating an experience that feels uncannily bespoke and acutely attuned to your emotional state.

Despite the progress, the journey to perfect content recommendation systems is ongoing. Transparency, collaboration, and responsible AI development will be key to ensuring these systems continue to enrich lives and inspire positive societal change. As platforms refine their algorithms and expand their data sets, the ultimate promise is to craft an entertainment landscape that is inclusive, engaging, and infinitely inspiring, aligning with both individual preferences and shared human values.

AI-driven Content Creation

In the ever-evolving landscape of entertainment, AI-driven content creation stands out as a transformative force, reshaping the way narratives are crafted and consumed. This section explores how artificial intelligence has emerged as a potent tool in creative industries, enabling new forms of storytelling and pushing the boundaries of artistic expression.

Artificial intelligence has long been associated with automation and efficiency, but its role in the creative process marks a new frontier. AI algorithms are now capable of generating music, writing scripts, and even producing visual art. This isn't just about speed; it's about exploring creative avenues that were previously uncharted. These AI systems, through machine learning and neural networks, can analyze vast datasets to recognize patterns and mimic human creativity. The results are often surprising and, at times, indistinguishable from human-produced content.

One of the most profound impacts of AI in content creation is its democratizing power. Aspiring creators, who may not have the resources of major production studios, can leverage AI tools to bring their visions to life. These tools enable creators to experiment with new ideas without the looming costs of traditional production, thus fostering innovation and inclusivity in the industry. By lowering the barriers to entry, AI-driven content creation allows diverse voices to flourish, contributing to a richer cultural landscape.

Moreover, AI has the ability to revolutionize storytelling by introducing adaptive narratives. Imagine a story that changes its plot based on the preferences and reactions of the audience. With AI-generated content, this dynamic interaction becomes possible, creating personalized experiences for each viewer or reader. This not only enhances engagement but also deepens the connection between the audience and the story, making content consumption a more interactive and immersive experience.

The integration of AI in content creation also has significant implications for traditional roles within the entertainment industry. Writers, musicians, and artists might initially feel threatened by the efficiency and capabilities of AI systems. However, rather than replacing human creativity, AI serves as a collaborator, assisting creators in developing new work. Through this collaborative process, creators can access unique insights and inspiration, using AI to enhance their creativity and efficiency.

Creativity, once considered the quintessence of human uniqueness, is now shared with machines, prompting philosophical reflections on the nature of creativity itself. Is a piece of art less valuable if it is produced by an AI? While AI can replicate styles and techniques, the human touch—empathy, emotion, and subjective experience—still holds a special place in art. The true potential of AI-driven content lies in its ability to augment rather than replace human creativity, enabling new forms of expression and innovation.

Beyond the creative process, AI also plays a critical role in managing and distributing content. Intelligent algorithms make content discovery both efficient and personalized, helping audiences find stories that resonate with them. In an era where content overload is prevalent, these recommendation systems ensure that niche and independent creations reach their potential audience, breaking the monopoly of mainstream media.

However, the rise of AI-driven content creation does come with challenges. There are ethical concerns regarding authorship and originality, as AI can easily replicate existing works, leading to issues of plagiarism and copyright infringement. Additionally, there is a risk of homogenization, where AI might cater to popular trends at the expense of originality, leading to a lack of diversity in narratives and cultural expressions.

In navigating these challenges, it is crucial to develop robust frameworks that ensure ethical use of AI in content creation. This involves establishing clear guidelines for authorship and attribution, as well as fostering transparency in the algorithms that drive content generation. Policymakers, industry leaders, and creators must collaborate to create a future where AI is harnessed responsibly to enhance creativity while respecting cultural heritage and intellectual property rights.

Looking ahead, the possibilities offered by AI-driven content creation are boundless. Virtual reality and augmented reality experiences, powered by AI, are poised to offer audiences a new level of immersion. Moreover, the development of generative AI, which can produce coherent and contextually relevant content, signals a future where AI could become an integral part of content studios and creative agencies, working alongside humans as co-creators of art and entertainment.

As we reflect on the role of AI in the creative industry, it becomes evident that we are at the threshold of a new era. AI-driven content creation has the potential to foster an explosion of creativity, innovation, and cultural exploration. By embracing this technology while addressing its ethical implications, we can ensure a future where AI and human creativity coexist, creating a vibrant and dynamic entertainment landscape that reflects the diversity and richness of human thought.

Chapter 14:
The Future of Work with AI Agents

As we step into a future sculpted by technological innovation, AI agents are set to redefine the workplace, evolving the nature of jobs and career landscapes. These intelligent systems are adept at automating routine tasks, which once consumed countless hours of human effort, freeing individuals to focus on work that requires creativity, critical thinking, and human touch. This transition doesn't mean jobs will vanish; rather, it signals a shift toward roles that blend human creativity with AI's analytic power. As businesses harness AI agents, we're seeing a burgeoning of new opportunities in fields like AI maintenance, ethics consultation, and hybrid roles where people and AI collaborate seamlessly. The future promises a dynamic workforce where learning and adaptation are key, turning potential challenges into avenues for growth and innovation. With ongoing advancements, this synergy between humans and AI agents may spearhead unprecedented productivity and creativity across industries, fostering a future where work not only evolves but flourishes.

Automation of Routine Tasks

The concept of automating routine tasks with AI agents is not just a futuristic vision—it's a tangible reality, transforming how industries and individuals approach work. At the heart of this transformation lies the unique capability of AI to handle repetitive and mundane tasks with unmatched efficiency. Machines, unlike humans, don't tire or lose focus, granting them the ability to tackle processes that were once considered the necessary but dull aspects of any profession.

Consider the laborious process of data entry. For years, employees have spent countless hours inputting information into databases, a necessity for maintaining accurate and up-to-date records. With AI agents, this task has become almost seamless. These agents can sift through vast amounts of data, identify relevant information, and autonomously update records. Not only does this save time, but it also reduces the margin for human error, thus enhancing the overall quality of data management.

In the world of finance, AI agents are making waves by automating tasks such as transaction categorization and customer query resolution. Financial institutions rely heavily on data precision, and even a minor mistake can lead to major ramifications. With AI, routine processing steps can be completed in a fraction of the time it previously took, with greater accuracy and reliability. This shift allows financial professionals to focus more on strategic tasks, such as risk analysis and investment planning, which are less susceptible to automation.

AI agents are also proving invaluable in the customer service domain. Traditionally, customer support centers required a significant workforce to manage inquiries, handle complaints, and process refunds. Implementing AI-driven chatbots and virtual assistants can dramatically alter this landscape. These intelligent agents not only understand and respond to customer inquiries in natural language but can also learn from past interactions to continuously improve their support quality. This results in faster response times and enhances customer satisfaction—a critical factor in today's competitive market.

The industrial sector is no stranger to automation, with robotics taking center stage in assembly lines. What AI agents offer to this sector is a new level of cognitive automation, enabling machines to adapt to changes and make decisions in real-time. This is particularly evident in quality assurance processes, where AI algorithms can detect defects or irregularities in products far more rapidly than the human eye can. Such capabilities ensure that only the highest quality goods leave the factory floor, which in turn strengthens brand reputation and customer trust.

While the efficiency gains are evident, it's essential to acknowledge that the shift towards automating routine tasks isn't without its challenges. Concerns around job displacement loom large as AI systems take on roles traditionally filled by human workers. However, history teaches us that technological advancements often lead to the creation of new opportunities. By automating routine tasks, we free up human resources to tackle more complex, creative, and strategic roles that machines can't yet replicate.

This transformation also underscores the pressing need for workforce adaptation. With AI handling the mundane, employees have the opportunity—and indeed the necessity—to upskill and reskill. Lifelong learning becomes a critical part of one's career as workers seek new ways to add value to their organizations. Educational institutions and corporations alike are recognizing this shift, offering programs that embrace the changing landscape of work.

The potential of AI to perform routine tasks extends even into the healthcare industry. AI agents are tasked with scheduling appointments, sending reminders, and even filing insurance claims. By taking these burdens off healthcare professionals, AI allows them to dedicate more time to patient care. Additionally, AI-driven systems can analyze patient data to identify trends and suggest preventative measures, further enhancing the quality of healthcare delivery.

In educational settings, AI can manage administrative duties such as grading assignments or tracking student attendance and performance metrics. This enables educators to focus more on teaching and interacting with students. By taking over the repetitive, time-consuming tasks, AI allows educational professionals to innovate and enrich the learning experience, potentially leading to improved educational outcomes.

Looking towards the future, the ability of AI to automate routine tasks will only expand, integrating more deeply into sectors we haven't even considered. With advancements in machine learning and natural language processing, AI's scope will widen, enabling these agents to undertake even more sophisticated roles that require understanding and contextual decision-making. This evolution won't be a simple replacement of human effort but a reimagining of what humans can achieve with the aid of intelligent systems.

There's no denying the impact AI agents are having on today's work environment. By offloading routine tasks, these systems unlock new efficiencies and possibilities, reshaping roles across industries. The narrative isn't just about machines replacing humans but how, together, they can create a more dynamic, effective, and innovative workforce. As AI continues to weave itself into the fabric of our daily lives, we'll need to adapt, embrace change, and remain inspired by the endless opportunities these technological advances present.

New Opportunities and Job Roles

As we stand at the cusp of a new technological era, AI agents are rapidly reshaping the landscape of work, creating opportunities that once seemed like science fiction. The integration of AI in various sectors doesn't just revolutionize existing roles—it also opens up an array of new career possibilities that are both exciting and challenging. From sectors as diversified as healthcare, finance, and entertainment to education and agriculture, the demand for a workforce equipped to harness AI's potential is unprecedented.

One of the most significant areas of opportunity lies in the realm of AI development and programming. As businesses strive to implement AI-driven solutions, there's a growing need for skilled professionals who can design, develop, and maintain these systems. Roles such as AI engineers, machine learning specialists, and data scientists are becoming mainstream. These individuals need to not only possess technical expertise but also an understanding of the ethical implications and business applications of AI technologies.

Beyond the technical roles, there's a burgeoning demand for professionals who can bridge the gap between technology and its practical applications. For instance, AI strategists and business analysts play a crucial role in identifying potential AI projects. They evaluate how AI can be integrated into existing systems to improve efficiency and innovation. Similarly, project managers with expertise in overseeing AI initiatives are essential, ensuring that projects are delivered on time, within budget, and to specification.

In the context of AI's interaction with human users, the demand for UX designers and specialists with AI-focused skills is also on the rise. These roles involve creating intuitive interfaces and ensuring that AI applications are as accessible as possible. Given the importance of user experience in AI applications, professionals who can effectively anticipate and respond to user needs are vital in ensuring that AI systems are not only efficient but also user-friendly.

The growth of AI is not limited to technical fields. As AI systems take on more complex tasks, ethical and policy considerations have become paramount. This has led to the emergence of roles in AI ethics and governance. These positions demand a deep understanding of both the technology and the societal impacts, ensuring that AI development and deployment adhere to legal, ethical, and human rights standards.

Another intriguing development is the rise of AI in creative fields, leading to hybrid job roles that weren't conceivable a few years ago. For example, AI specialists are now collaborating with artists to create generative art, music, and even literature. Roles that combine technical AI expertise with artistic flair are increasingly favored, driving a new wave of innovation in creative domains.

Moreover, the education sector is witnessing transformative job roles as AI-driven personalized learning systems become more prevalent. Teachers and educators are exploring roles that focus on instructional design and tech integration. There's a growing recognition of the need to adapt teaching methods to include AI tools, which not only enhance learning but also personalize educational experiences for students.

The healthcare industry is experiencing a paradigm shift with the emergence of AI-assisted diagnosis and treatment planning. This has created insights-driven roles such as healthcare data analysts and AI-assisted diagnostic professionals. They are tasked with interpreting AI-generated data to support medical decisions. Additionally, medical practitioners are now trained in AI literacy, learning to work alongside AI in diagnosing and treating patients more effectively.

Meanwhile, AI is revolutionizing traditional fields such as finance, where roles in automated trading and fraud detection are expanding. Finance professionals now need to be equipped with AI skills to analyze large datasets, predict market trends, and enhance security measures. This blend of financial acumen with AI proficiency is creating new career trajectories and growth paths within the industry.

The realm of cybersecurity is also undergoing a transformation due to AI advancements. As threats become more sophisticated, cybersecurity experts with AI skills are becoming indispensable to protect data and systems from complex cyber attacks. Roles that combine cybersecurity with AI applications are therefore on the rise, underscoring the critical role AI plays in safeguarding information.

In manufacturing and industrial sectors, AI's influence is creating jobs in orchestrating smart factories and managing automated processes. Engineers and technicians need to build and maintain AI-driven machinery and systems, focusing on efficiencies previously unattainable without AI's insights and capabilities. These roles often require a blend of engineering expertise and knowledge of AI's capabilities.

Furthermore, AI's integration into everyday life is escalating the need for AI product managers and sales specialists who can market AI solutions effectively. These roles involve understanding complex AI products and conveying their value to businesses and consumers. The blend of technical knowledge and sales experience is crucial in driving adoption in the competitive marketplace.

Finally, AI's impact on energy and environmental sectors can't be overlooked. As AI becomes instrumental in optimizing energy systems and monitoring environmental changes, new positions are arising in AI-enhanced environmental management and renewable energy systems. Professionals in these fields are tasked with leveraging AI to improve sustainability efforts, which is critical for tackling climate change and other global challenges.

In conclusion, while AI roles span numerous fields, the key driver is adapting skills to align with technological advancements. The focus should be on continuous learning, cross-disciplinary knowledge, and understanding AI not just as a tool but as a partner in innovation. As AI continues to evolve and integrate into diverse sectors, embracing its potential to create new opportunities and roles is vital. The future of work isn't merely about replacing human jobs with machines—it's about creating a symbiosis where human ingenuity and AI capabilities merge to define the next chapter in human enterprise.

Chapter 15:
AI Agents in Research
and Development

In the fast-paced realm of research and development, AI agents are becoming indispensable allies, propelling advancements that once seemed out of reach. They accelerate scientific discoveries by analyzing vast datasets with unmatched speed and precision, unveiling patterns and insights that often elude human observation. These agents empower researchers to innovate with greater accuracy, from decoding complex genomic structures to simulating sophisticated chemical reactions. By handling labor-intensive tasks, they free researchers to focus on creative problem-solving and strategic thinking. AI agents also foster collaboration across disciplines, enabling the synthesis of diverse knowledge streams into groundbreaking solutions. In the journey of discovery and innovation, these intelligent partners are not just tools, but catalysts, forging a new era of exploration and possibility in science and technology.

Accelerating Scientific Discoveries

The landscape of scientific research has always been defined by its relentless pursuit of knowledge and innovation. In recent years, however, artificial intelligence (AI) agents have emerged as transformative tools, not just by speeding up the pace of discoveries but also by opening new avenues of inquiry that were previously inconceivable. By automating routine tasks, identifying patterns, and making accurate predictions, AI agents are revolutionizing how scientists conduct research, analyze data, and test hypotheses.

The hallmark of modern scientific exploration lies in the sheer volume of data generated. From genomics to astrophysics, the datasets are enormous, creating a bottleneck for traditional analysis methods. Here, AI agents shine by leveraging machine learning algorithms capable of processing vast amounts of data with unparalleled speed and precision. This capability allows researchers to move beyond descriptive analyses, decipher complex relationships within data, and uncover hidden patterns that might elude the human eye.

One notable area where AI agents are making substantial contributions is drug discovery. The traditional pipeline for bringing a new drug to market is time-consuming and expensive. However, AI agents can sift through chemical databases, predict molecular interactions, and even suggest potential compounds for further testing. This not only accelerates the initial stages of drug development but also improves the likelihood of finding viable candidates, particularly in areas of urgent medical need such as antibiotic resistance and rapidly evolving viruses.

AI agents have also infiltrated laboratory environments, particularly through the deployment of automation and robotics. These intelligent machines can perform repetitive tasks with high precision, allowing human researchers to focus on creative and interpretive aspects of their work. For example, AI-driven robots can conduct high-throughput screening of biological samples or carry out complex experimental procedures around the clock, thereby significantly reducing the time needed for experimental verification.

In fields such as material science, AI agents assist by predicting the properties of new materials before they are synthesized. By simulating and modeling the physical behaviors of materials at atomic and molecular levels, AI can help identify promising candidates with desirable attributes like strength, conductivity, or biocompatibility. This streamlines the research cycle, enabling rapid testing and adaptation, which is a boon for industries aiming to develop sustainable and advanced materials.

Moreover, AI is redefining the theoretical realms of science. The vast computational power of AI systems allows scientists to tackle complex problems in physics, chemistry, and other disciplines through simulations that were previously impossible. By providing novel insights into phenomena like quantum mechanics and cosmology, AI not only accelerates discovery but occasionally redefines the questions researchers are asking.

AI agents are also beginning to play a role in collaborative research, connecting disparate fields and promoting interdisciplinary studies. By analyzing data from multiple sources, AI can suggest correlations or areas of interest that bridge different scientific domains. This catalyzes transformative discoveries, fostering a deeper understanding of complex systems and processes—ranging from ecosystems to the human body—that rely on the integration of various scientific perspectives.

However, the integration of AI in scientific research isn't without its challenges. There are concerns regarding the interpretability of AI models, particularly in critical fields like healthcare and public safety. Ensuring these models are transparent and explainable is crucial for building trust and ensuring ethical implications are considered. Researchers must balance relying on AI outputs with their discernment and expertise, ensuring that AI serves as a tool and not a crutch.

Equally important is the training and preparation of researchers to work alongside AI agents. As AI tools become more sophisticated, scientists must acquire the skills to leverage these technologies effectively. This calls for an education paradigm shift, incorporating data science and AI literacy into the core training of future scientists and researchers to maximize the potential of these technologies.

As we stand on the cusp of an era brimming with potential, AI agents are pivotal players in accelerating scientific discoveries. They empower researchers to push the boundaries of what is known and venture into uncharted territories with confidence. The fusion of human ingenuity and AI's raw computational power is a partnership that could unravel mysteries that have puzzled humanity for centuries, paving the way for innovations that were once the realm of science fiction. By embracing AI, the scientific community is poised to not just write a new chapter but to redefine the very nature of discovery itself.

Advanced Data Analysis Techniques

As AI agents carve out a niche within research and development, their impact on data analysis is nothing short of transformative. Advanced data analysis techniques, powered by AI, are equipping researchers with tools and insights that were once unimaginable. These techniques not only enhance the efficiency of data processing but also redefine the boundaries of what data can reveal. In today's rapidly evolving research landscape, leveraging such techniques is no longer optional but essential for uncovering profound insights.

At the core of these advanced techniques lies machine learning, particularly deep learning. This subset of AI, with its ability to process vast datasets, enables researchers to identify patterns and trends that are invisible to the human eye. Deep learning models, trained on diverse datasets, can perform complex tasks such as image and speech recognition with remarkable accuracy. They allow researchers to automate the extraction of data insights, reducing the time between data collection and actionable intelligence.

Consider the role of AI in genomics. The analysis of genetic data is a daunting task, given the sheer volume of information encoded within DNA sequences. AI agents, through advanced data analysis techniques, are enabling breakthroughs in genomics by facilitating the rapid analysis of genetic variations. This acceleration aids in the identification of potential genetic biomarkers for diseases, paving the way for personalized medicine.

Furthermore, AI-driven data analysis is proving invaluable in the realm of climate science. Here, AI agents process extensive datasets from various environmental sensors and satellites to model climate patterns. By harnessing machine learning algorithms, researchers can make more accurate climate predictions and better understand the impacts of climate change. This not only aids policymakers but also empowers scientists to develop strategies aimed at mitigating adverse climate effects.

Another fascinating application of these techniques is in the field of social sciences. Traditionally reliant on qualitative analysis, social sciences are now benefitting from AI agents' ability to quantitatively analyze complex human behavior patterns. Through natural language processing (NLP) and sentiment analysis, AI systems can interpret vast amounts of social media data to gauge public opinion or predict social phenomena. Such insights are proving invaluable for governments, businesses, and researchers alike.

The richness of AI-enhanced data analysis isn't confined to just uncovering trends; it's also about enabling discovery that seems serendipitous. Often, AI algorithms identify correlations or anomalies that wouldn't have been apparent through conventional analysis techniques. This capability can drive scientific discoveries, leading to new lines of inquiry or even entire research paradigms.

However, deploying AI in advanced data analysis isn't without its challenges. One significant concern is ensuring the consistency and quality of the data fed into AI models. Poor data quality can lead to inaccurate results, potentially derailing research projects. As such, data preprocessing is critical, often involving cleaning, normalization, and augmentation of data to make it suitable for analysis.

Moreover, interpretability of AI models remains a topic of intense debate and research. While AI systems can provide highly accurate predictions, understanding how these results are derived is equally important, especially in fields like medicine, where accountability is crucial. Researchers are striving to develop models that not only predict outcomes but can also explain their decision-making process in human-understandable terms.

The future of advanced data analysis will likely see a blend of AI-driven techniques and traditional methods. Hybrid models combining statistical methods with machine learning are emerging, offering the best of both worlds. These models hold promise in delivering robust, accurate, and interpretable insights across various domains.

Indeed, the capabilities of AI in data analysis can be seen as both a catalyst and a compass. They catalyze the speed and scope of research while guiding researchers along the most promising paths of inquiry. As AI agents continue to evolve, their impact on the research and development landscape will only grow, highlighting the importance of investing in AI technologies.

In conclusion, advanced data analysis techniques powered by AI agents are redefining research and development. By enabling faster and more accurate analysis, these techniques empower researchers to tackle complex challenges and unlock new opportunities. As AI continues to mature, its role in data analysis will undoubtedly expand, shaping the future of research for years to come.

Chapter 16:
AI Agents in Smart Homes

As artificial intelligence increasingly intertwines with daily life, the concept of smart homes has morphed from science fiction into a reality reshaping our day-to-day experiences. AI agents are the backbone of this transformation, providing seamless automation and intelligence within our living spaces. These digital companions manage everything from adjusting thermostats to optimizing lighting and enhancing energy efficiency, all tailored to the homeowner's lifestyle and preferences. The magic lies in their ability to learn and adapt, offering solutions that not only enhance convenience but also promote sustainability by minimizing energy wastage. Imagine waking up to an AI-curated ambience that greets you with your favorite morning playlist, opens blinds as the sun rises, and ensures your coffee is brewing as you step into the kitchen. This harmonious integration of AI in smart homes represents a bold step toward intelligent automation, promising an upgraded quality of life while advocating for a more eco-friendly living environment. As these agents evolve, they hold the potential to address more complex challenges, bringing us closer to a future where our homes intuitively respond to our needs, blending technology with everyday living in ways previously unimaginable.

Home Automation Systems

Picture waking up each morning to soothing music that nudges you gently from slumber, while the aroma of freshly brewed coffee wafts in from the kitchen. Your smart thermostat adjusts the room temperature to your preferred setting as you step into a bathroom where the lights intensify from a dim glow to the perfect brightness for your morning routine. All these seemingly orchestrated actions are the result of comprehensive home automation systems powered by AI agents, working tirelessly in the background to elevate comfort and convenience in your everyday life.

Home automation systems represent a significant leap forward in integrating AI technology into residential settings. At their core, these systems aim to create a seamless and responsive environment where technology anticipates and meets the needs of the inhabitants without direct intervention. AI agents play a pivotal role in this transformation, functioning as the brains behind the smart home environment, enabling devices to communicate, adapt, and optimize operations for efficiency and personalization.

The backbone of these sophisticated systems is a blend of IoT devices, connectivity protocols, and AI algorithms. IoT devices within homes, ranging from smart speakers and cameras to connected kitchen appliances, collect and transmit data through various protocols. This data is then processed using AI to make intelligent decisions that streamline home operations. For instance, AI can analyze patterns in energy consumption, suggesting optimal heating schedules to enhance energy efficiency without compromising comfort.

In the realm of home security, AI agents serve as vigilant guardians. They can process visual and audio data from security cameras and sensors, identifying potential threats based on learned patterns and anomalies in real-time. This proactive monitoring not only heightens security but also provides peace of mind, as homeowners receive alerts and can review footage or incidents directly from their mobile devices, wherever they are.

AI-driven home automation systems also excel in tailoring personalized experiences. Consider lighting systems that adjust tones and intensities based on the time of day, activities, or even the mood preferences of individuals. Such systems utilize machine learning algorithms to learn and adapt to the behaviors and preferences of residents, creating a truly bespoke ambiance cost-effectively.

The interaction with these smart systems primarily happens through digital assistants embedded within devices like smart speakers or smartphone apps. More than mere conduits for commands, these assistants utilize natural language processing (NLP) to understand, interpret, and even anticipate user needs. This hands-free, voice-operated interaction enables residents to control various aspects of their home environment with ease and precision.

Privacy considerations, however, become paramount as AI systems gain access to personal data and daily routines. Users and developers alike face the critical task of safeguarding this information against breaches and misuse. Balancing privacy with convenience demands robust security protocols and transparency about data usage, underscoring the ongoing dialogue between technological advancement and ethical standards.

Moreover, the success of AI in home automation lies not only in the technology itself but also in user trust and acceptance. For the technology to thrive, individuals must feel confident that these systems will enhance their lives without exerting unwanted control or complexity. Ensuring easy-to-use interfaces and reliable performance are key factors in fostering this trust and catalyzing widespread adoption.

In conclusion, home automation systems signify much more than technological convenience—they herald a fundamental shift towards intelligent living environments that enhance quality of life. As AI agents grow more sophisticated, their potential to transform how we interact with our homes and surroundings continues to expand, unlocking new realms of possibility and innovation.

Enhancing Energy Efficiency

In the tapestry of AI's integration into our lives, enhancing energy efficiency within smart homes emerges as a compelling narrative. This is a domain where AI agents truly demonstrate their capacity to revolutionize daily living, making our homes not just more intelligent, but sustainably smarter. The intersection of AI and energy efficiency in homes is not simply about reducing costs; it's about pioneering a future where energy usage is intuitive, responsive, and remarkably reduced.

Imagine a smart home where energy consumption doesn't operate under a one-size-fits-all model, but is meticulously tailored to the habits and preferences of its occupants. AI agents achieve this by learning from patterns of energy usage and making informed decisions to optimize consumption. For instance, they can adjust heating or cooling systems based on occupancy data and ambient conditions, ensuring maximum comfort with minimal waste. These intelligent adjustments might seem small in the moment, yet they accumulate to significant savings and reduced carbon footprints over time.

Central to this optimization is the ability of AI agents to process and analyze vast datasets, seizing insights that elude human perception. They continually gather data from sensors deployed throughout the smart home, piecing together a dynamic portrait of energy flow. This perceptiveness enables AI to identify inefficient energy usage, adapting the system to eliminate wastage. The potential benefits echo both in the wallet and in the wider environmental landscape.

Moreover, AI agents empower homeowners with unprecedented autonomy and control over their energy usage. Interactive dashboards and intuitive user interfaces allow homeowners to monitor energy consumption in real time, diving into detailed reports about how energy is used at different times of the day and across various devices. With this information, households can make informed decisions, engage in budgeting exercises, and even set custom goals for sustainability efforts. AI agents make energy management a participatory experience, galvanizing residents to be proactive in their energy consumption.

In more advanced implementations, AI agents manage distributed energy resources such as solar panels or home batteries, orchestrating the balance between energy production and consumption. They can forecast energy generation based on weather patterns and schedule the charging or discharging of batteries accordingly. This capability not only optimizes household energy independence but can also support grid stability, allowing homes to contribute actively to broader energy networks.

The implications of these technologies extend beyond the immediate sphere of the domestic environment. By fostering widespread adoption and improvement in energy efficiency, AI agents contribute significantly to global efforts in tackling climate change. They represent a pivotal shift towards integrating renewable energy sources more seamlessly into everyday life, reducing reliance on non-renewable resources, and minimizing our overall ecological footprint.

Yet, the journey towards fully optimized smart homes is not without its hurdles. Challenges such as data privacy and security need careful navigation. While AI agents harness data for energy efficiency, the assurance of data protection is paramount. Smart home systems must incorporate robust encryption and secure protocols, ensuring that sensitive user data isn't accessible for malicious intent. Moreover, homeowners must be educated about these technologies, fostering trust and understanding in how AI aids in their energy efficiency goals.

Equally important is ensuring interoperability among various smart home devices and systems. The diverse ecosystem of smart devices should seamlessly communicate with AI agents, avoiding siloed solutions that limit efficiency. Industry standards and collaboration among tech companies will be vital in overcoming these interoperability challenges, ensuring AI solutions can integrate easily within existing infrastructures.

Looking forward, the future of AI in enhancing energy efficiency is laden with potential. As AI technologies advance, we anticipate even greater granularity in understanding and controlling energy resources. Machine learning algorithms will become even more adept at predicting usage patterns and anomalies, providing actionable insights that lead to a new level of precision in energy management.

Moreover, as AI agents evolve, so will their role as educators and advisors. They might not only manage energy efficiency but also educate occupants on sustainable practices, suggesting ways to minimize consumption or highlighting the environmental impacts of their choices. This capacity to inspire behavioral change adds another dimension to their utility in achieving global sustainability targets.

Ultimately, AI agents in smart homes are not just about the immediate gains of saving energy and costs; they are about carving a path towards a greener, more sustainable future. Through intelligent decision-making, AI is transforming our living spaces into ecosystems that harmonize comfort with conservation. The prospect of homes that intuitively conserve energy speaks directly to our shared aspirations for living in balance with the environment and illustrates the profound potential that AI holds in reshaping how we interact with our world.

Chapter 17:
Transforming Agriculture
with AI Agents

In today's rapidly evolving technological landscape, the agricultural sector stands on the brink of a revolution, driven by AI agents that promise to transform traditional practices into highly efficient systems. Precision farming techniques are at the forefront, using AI to analyze vast amounts of data for enhanced soil management and optimal crop yield predictions. Farmers are no longer solely reliant on intuition; AI-driven crop monitoring systems provide real-time insights on plant health, water usage, and pest infestations, empowering faster decision-making and resource conservation. These AI agents not only maximize productivity and sustainability but also offer hope in addressing the global challenge of food security. This monumental shift heralds a new era of agriculture, where the fusion of technology and nature creates farming practices that are both intelligent and sustainable, meeting the needs of an ever-growing population while preserving the planet's precious resources.

Precision Farming Techniques

In the evolving landscape of agriculture, precision farming stands as a beacon of innovation, enhancing the efficiency and productivity of food production. AI agents are at the forefront of this transformation, integrating advanced technologies and methodologies into traditional farming practices. As agricultural demands increase, precision farming techniques offer sustainable solutions, reducing waste and optimizing resources. This integration allows farmers to apply inputs more precisely in terms of amount, timing, and location, maximizing yields while minimizing environmental impact.

At the heart of precision farming are AI-driven tools that analyze vast amounts of data to offer insights and make decisions. Satellite imagery and drones, equipped with multispectral and hyperspectral cameras, capture comprehensive field data. These images are processed by machine learning algorithms that identify patterns and anomalies in crop health, soil quality, and water distribution. With this information, AI systems can generate detailed maps that inform farmers where to focus their efforts.

One of the critical components of precision farming is the use of soil sensors combined with AI analytics. These sensors collect data on moisture levels, temperature, and nutrient content in real-time. By integrating this information, AI models predict the best times for planting and harvesting, tailored to each specific plot of land. This adaptation reduces the likelihood of crop failure and ensures that resources such as water and fertilizers are used efficiently. Consequently, farmers can achieve higher productivity with less environmental strain.

Variable Rate Technology (VRT) exemplifies precision farming, where AI-driven automation adjusts the rate of application of seeds, fertilizers, and pest control agents based on the condition of individual farm areas. This technique enhances crop yield by investing the right resources only where they are needed. The elegance of VRT is in its adaptive nature; AI systems learn from past applications and continuously refine their recommendations, creating a dynamic farming environment that responds to real-time conditions.

Precision irrigation systems, steered by AI, further underscore the transformative power of technology in agriculture. These systems analyze weather forecasts, soil moisture, and plant water needs to optimize irrigation schedules. They can automate water dispersal to specific sections of the farmland, saving water and ensuring plants receive exactly what they need, when they need it. This targeted approach diminishes water waste and improves crop health, which is vital in regions suffering from water scarcity.

Another pivotal technique is crop monitoring with AI-enabled pest and disease detection. By leveraging image recognition and natural language processing, AI systems assess plant health through visual and verbal data. This capability allows farmers to detect early signs of pests or diseases, facilitating timely interventions that prevent widespread damage. The automation of monitoring tasks frees up farmers to concentrate on strategic decision-making rather than routine field checks.

Incorporating precision farming techniques requires not just technological adoption but also a shift in mindset. Farmers need to become data literate, embracing a digital approach to agriculture. Training and support systems offered by tech companies ensure that farmers understand how to use AI tools effectively. This skill enhancement is crucial for maximizing the benefits that precision farming offers, fostering a symbiotic relationship between human expertise and machine efficiency.

Beyond individual farms, precision farming techniques contribute to food security on a larger scale by optimizing supply chains. AI systems forecast yield outputs, helping supply chain stakeholders improve their planning and reduce waste. This predictive capability ensures that the right quantities of produce reach markets at the right time, balancing supply and demand efficiently.

Precision farming's environmental benefits can't be overstated. By ensuring precise input usage, these techniques reduce the agricultural footprint on ecosystems, preserving biodiversity and minimizing runoff that can leach into water sources. AI agents thus become allies in the quest for sustainability, lowering agriculture's carbon footprint and conserving resources for future generations.

The future of precision farming lies in its potential for scalability and adaptability. As AI models gather more data across diverse agricultural landscapes, their accuracy and reliability will only increase. The integration of IoT devices and AI agents into every aspect of farming—planting, monitoring, and harvesting—will drive unprecedented efficiencies, paving the way for a new era in agriculture.

In conclusion, precision farming represents a synthesis of tradition and technology, where AI agents empower farmers with the tools to make informed, data-driven decisions. By harnessing the capabilities of AI, precision farming techniques transform challenges into opportunities, driving the agricultural sector toward a future that is both resilient and prosperous. The ripple effects of these advancements promise to nourish a growing population while safeguarding the planet's resources, embodying a vision of sustainability and innovation. As AI continues to evolve, its role in agriculture will undoubtedly deepen, leading to even more sophisticated solutions and a bountiful tomorrow.

AI-driven Crop Monitoring

AI-driven crop monitoring stands at the forefront of agricultural innovation, offering unprecedented precision that is transforming farming practices across the globe. By harnessing the power of machine learning algorithms, farmers can now gain insights into crop growth and health at microscopic levels, something that was once considered science fiction. This technology is a critical component of precision farming, enabling cultivators to maximize yields while minimizing waste and environmental impact.

The advent of AI in agriculture is a game changer, integrating diverse data sources such as satellite imagery, drone surveillance, weather patterns, and soil sensors to provide farmers with detailed, real-time updates on their crops. These insights are vital for making informed decisions about irrigation, pest control, and fertilization. When farmers know exactly what their crops need and when they need it, the efficiency of these processes skyrockets.

For instance, AI systems can predict pest infestations before they become visible, allowing farmers to apply tailored interventions that are both cost-effectively and environmentally sound. Traditional methods often resulted in overuse of pesticides and other chemicals, which could harm the ecosystem. AI-driven solutions propose a sustainable alternative by assessing risk factors and suggesting targeted responses.

Consider how intricate machine learning algorithms process data collected from sensors placed in fields. These sensors measure soil moisture, nutrient levels, and even the pH balance, all crucial indicators of crop health. With AI, this complex information is transformed into actionable insights, guiding the farmer to apply interventions precisely where and when they are needed. This level of detail ensures that resources such as water and fertilizer are used optimally, conserving natural resources and reducing costs.

Drones, equipped with advanced camera systems, hover over fields capturing detailed imagery of crop yields and potential problem areas. These high-resolution images are then analyzed by AI programs to detect issues like plant diseases or nutrient deficiencies. The transformative aspect of this technology lies in its ability to cover vast expanses of farmland quickly and provide insights that humans would struggle to gather manually.

Weather prediction is another critical area where AI-driven crop monitoring shows its advantage. By analyzing weather patterns and forecasts, AI can anticipate droughts, frosts, or storms, providing farmers with the lead time needed to protect their crops. Access to this predictive modeling significantly shifts farming from a reactive to a proactive industry, vital in an age of climate change and unpredictable weather patterns.

Moreover, these technologies advocate for sustainable practices by promoting no-till farming and crop rotation based on data insights. AI systems can offer suggestions on ideal planting times and crop combinations that can enhance soil health rather than deplete it. This ensures ongoing productivity while maintaining ecological balance.

Adoption of AI-driven crop monitoring is gaining momentum worldwide, yet it also presents unique challenges. Rural areas, which are at the heart of agriculture, often suffer from limited access to high-speed internet, impeding the full deployment of technology solutions that depend on cloud computing. Moreover, the cost of implementing AI systems can be a barrier for small-scale farmers; however, emerging cost-effective solutions and government subsidies are making these technologies more accessible.

Data privacy remains a critical concern as well. Farmers need assurance that the immense amounts of data collected will be stored securely and utilized ethically. Establishing transparent policies around data usage and benefits will be key as AI integration becomes more widespread.

AI-driven crop monitoring isn't merely about advanced technology—it's about transforming agriculture into a sustainable, efficient industry. By combining continuous monitoring with machine learning, the agricultural sector can feed a growing global population while respecting the planet's ecological limits. This technology exemplifies the potential for AI to rise to the most pressing challenges of our time, paving the way for an era of smarter, more responsible agriculture.

The inspirational aspect of AI-driven crop monitoring lies in its potential to empower farmers, big and small. By placing critical insights into the hands of those who nurture the earth, AI offers the promise of better yields, higher profits, and a more sustainable approach to feeding the world. As this technology continues to evolve, its capacity to foster resilience in food systems will be pivotal, ensuring food security and sustainability for generations to come.

Chapter 18:
AI Agents in Environmental Management

As humanity grapples with mounting environmental challenges, AI agents emerge as powerful allies in orchestrating sustainable solutions. These intelligent systems are revolutionizing how we monitor and respond to climate change, offering unprecedented precision in data analysis and forecasting. By capturing real-time information from an intricate web of sensors and satellites, AI agents help predict weather patterns, track greenhouse gas emissions, and evaluate the health of ecosystems with remarkable accuracy. In conservation efforts, these agents tirelessly analyze vast tracts of data to identify endangered species, monitor habitats, and even combat illegal poaching activities. This sort of agility transforms proactive environmental management from a daunting prospect into a tangible reality, inspiring both immediate and long-term strategies that align with ecological harmony. Through collaboration and innovation, AI agents are not just tools—they are key partners in our quest to safeguard the planet for future generations, forging pathways that redefine how we coalesce with nature.

Monitoring Climate Change

As our planet grapples with the escalating intricacies of climate dynamics, leveraging AI agents for monitoring purposes offers unprecedented potential. This transformative technology acts as a vigilant sentinel, deciphering the myriad data points related to climate change with remarkable speed and precision. But what does this mean for the future of environmental management?

AI agents play a critical role in collecting, processing, and interpreting data from various sources. Satellites equipped with advanced sensors continuously gather information about Earth's atmosphere, oceans, and land surfaces. These satellites produce massive datasets that AI algorithms adeptly analyze. This capability allows scientists to track climate patterns and anomalies more effectively than ever before, from global temperature fluctuations to regional precipitation changes.

The ability to predict climate-related phenomena with greater accuracy is perhaps one of the most exciting prospects. By harnessing machine learning models, AI agents improve forecasting methods for extreme weather events like hurricanes, floods, and droughts. These forecasts not only aid in preparation and response but also help mitigate the adverse impacts of such disasters on communities and ecosystems.

The deployment of AI agents extends beyond high-altitude satellites; terrestrial sensors and IoT devices are proliferating in environment-focused applications. From monitoring deforestation rates in the Amazon to assessing coral reef health in the Pacific, these devices accumulate granular data that AI can process in real-time. Through pattern recognition and anomaly detection, AI agents can alert scientists to subtle changes indicating potential environmental threats.

Incorporating AI technologies into climate research enhances traditional methods. When combined with computer simulations, AI agents refine climate models to reflect a more accurate picture of Earth's complex systems. This iterative approach reduces uncertainties in projections and informs policymakers as they craft climate adaptation and mitigation strategies.

Moreover, AI's role in monitoring climate change isn't confined to observation alone. It extends to proactive engagement. For instance, AI agents are instrumental in optimizing renewable energy infrastructure. They forecast energy demand, adapt to resource availability, and even control grid interactions, facilitating smoother transitions from fossil fuels to sustainable alternatives.

AI drives efficiency and precision in environmental conservation efforts. Machine learning techniques analyze data on species populations, migration patterns, and habitat changes, contributing to biodiversity preservation. These insights guide conservation strategies, enabling targeted actions that maximize ecological benefits.

The implications of AI-based climate monitoring aren't limited to scientific research and policy-making. They're impacting how organizations and businesses approach sustainable practices. Companies are increasingly adopting AI-driven analytics to assess their carbon footprints, identify inefficiencies, and monitor compliance with environmental regulations. By integrating AI insights, businesses can pursue sustainability goals with heightened accountability and impact.

Furthermore, AI's capacity to synthesize complex datasets fosters collaboration across disciplines. It brings together climatologists, data scientists, ecologists, and policy experts, fostering a multidisciplinary approach to tackling climate change. This collaborative synergy empowers stakeholders to devise novel solutions that leverage diverse expertise.

While AI offers formidable potential in climate monitoring, challenges remain. Data privacy, algorithmic transparency, and equitable access must be prioritized to ensure that AI benefits all.

Creating robust frameworks to manage these challenges will be paramount as we navigate the delicate balance between technological advancement and ethical responsibility.

Future advancements in AI agents will further enhance our ability to monitor and respond to climate change. Emerging technologies like edge computing promise more agile and localized data processing, reducing latency and increasing deployment possibilities in remote or underserved regions. This could democratize access to life-saving climate information globally.

In conclusion, AI agents in environmental management mark a paradigm shift in how we perceive and respond to climate change. They equip us with the tools to understand our planet's intricate systems more deeply and respond to its changes more sustainably. The road ahead in monitoring climate change with AI is not solely about observing—it is about empowering informed action, galvanizing innovation, and fostering resilience in the face of one of humanity's greatest challenges.

AI in Conservation Efforts

In a world increasingly burdened by environmental challenges, the role of artificial intelligence (AI) in conservation efforts is more vital than ever. AI agents have transformed how we approach the conservation of biodiversity, natural resources, and ecosystems. By providing real-time data and insights, AI empowers conservationists to make informed decisions, helping to safeguard our planet for future generations.

One of the most significant contributions of AI in conservation is its ability to process vast amounts of data. Remote sensing technology, including satellite and drone imagery, generates extensive datasets tracking habitat changes, deforestation, and wildlife movement. Machine learning algorithms sift through these data troves to identify patterns and anomalies that might escape human attention. This efficiency in data analysis enables faster response times to emerging conservation issues.

Moreover, AI agents are pivotal in species protection, particularly with endangered animals. Poaching and illegal wildlife trade pose significant threats to biodiversity. AI technologies like predictive analytics and pattern recognition can anticipate poaching activity by analyzing crime data and spotting trends that indicate illegal activity. Innovative solutions such as real-time alerts enable rangers and conservation organizations to deploy resources effectively and take preemptive actions to deter poaching.

AI exceeds its utility at preventing criminal activity, contributing significantly to species tracking and monitoring. Camera traps outfitted with AI algorithms can identify species automatically and log their movements without human oversight. This capability not only enhances data accuracy but also reduces the human resources required for species monitoring. Additionally, AI-powered acoustic sensors track elusive species through sound, such as the calls of whales or birds, uncovering the mysteries of their habitats and behaviors.

However, AI in conservation isn't just about solving problems. It also allows scientists to ask more profound questions about ecosystems and biodiversity. AI models can simulate complex ecological interactions, offering insights into how environmental changes can impact various species and their habitats. These simulations help in crafting robust conservation strategies adaptable to future climate scenarios and human activities.

Despite its potential, deploying AI in conservation faces challenges and limitations. There are ethical concerns regarding data privacy, as the collection of geospatial and census data may inadvertently infringe on the rights of local communities. AI solutions must be designed and integrated carefully, considering the socio-cultural contexts of the regions they aim to serve.

Another major hurdle is the technological gap between developing and developed regions. Many biodiversity hotspots are located in areas with limited access to advanced technology. Bridging this digital divide is crucial for enabling global conservation efforts, requiring investment in infrastructure, education, and capacity building to ensure that AI tools are accessible and usable across diverse settings.

Innovation in AI also prompts questions about the sustainability and ethics of the technologies used. The energy-intensive nature of AI systems can contribute to climate issues they aim to solve. Therefore, ensuring that AI solutions for conservation are energy-efficient and have low carbon footprints is paramount. Researchers are developing lightweight algorithms and harnessing renewable energy sources to power AI systems, aiming to align technological advancements with ecological sustainability.

Community involvement is another essential aspect when implementing AI in conservation. Engaging local populations and integrating indigenous knowledge with AI models can enhance the effectiveness and acceptance of technology-driven conservation measures. AI should not replace human intuition and experience but rather augment it, offering new tools and perspectives to empower communities to protect their natural heritage.

AI's role in conservation marks a promising frontier, potentially transforming how we understand and protect our natural world. As these technologies mature and become more inclusive, their contributions will likely expand, creating new possibilities for environmental stewardship. The integration of AI into conservation efforts embodies a union of technology and nature, illustrating a path forward where innovation serves the greater good of preserving life in all its forms.

Chapter 19:
Technical Challenges in AI Development

As AI continues to revolutionize industries, it faces a myriad of technical challenges that must be navigated with precision and innovation. Developing scalable AI systems is paramount as demand grows and applications expand, yet achieving this requires sophisticated infrastructure capable of evolving with technology itself. Overcoming computational limitations is another critical hurdle; with increasing complexity and data-centric processes, AI systems demand vast computational power, often straining even the most advanced hardware. Furthermore, refining algorithms to enhance efficiency without sacrificing performance requires a deep understanding of both the problem space and transformative techniques. Balancing these demands while fostering innovation is no small feat, but it is within these challenges that the potential for groundbreaking advancements lies. By addressing these obstacles, AI developers not only push the limits of what's achievable but also drive forward the future of intelligent automation, bringing us closer to a world where AI seamlessly integrates into daily life and unlocks unprecedented possibilities.

Scalability of AI Systems

The promise of artificial intelligence lies in its potential to operate at a scale that matches human creativity and problem-solving, but this promise poses significant technical challenges. Scalability in AI systems is not just about expanding data capacity or increasing processing power. It involves a complex interplay of technological, algorithmic, and infrastructural components that need to evolve concurrently. AI systems must be capable of handling vast amounts of data in real time while also making intelligent decisions based on that data. With the increasing demand for sophisticated AI applications across various domains, the challenge of scalability becomes more prominent.

In simple terms, scalability refers to an AI system's ability to grow and manage increased demand efficiently. For AI to take the leap from a handful of niche applications to widespread use, its frameworks must be scalable. This involves two primary aspects: vertical scalability, which entails enhancing the system within a single node, and horizontal scalability, which means expanding the AI system across multiple nodes or devices. Both aspects demand advancements in computational power, storage solutions, and network capabilities.

One of the notable challenges in scalability is the exponential growth of data. As AI agents become more embedded in our daily lives, the data they generate multiplies. Unstructured data, found in social media streams, IoT devices, and more, requires sophisticated handling. Efficient storage solutions, such as cloud-based systems and distributed databases, become integral in managing this deluge of information. These systems need to not only store vast amounts of data securely but also facilitate rapid retrieval and analysis.

Algorithms play a pivotal role in determining scalability as well. Machine learning models and neural networks must be designed to process massive datasets without compromising performance. As datasets become larger, the training times for these models can become prohibitively long, thereby necessitating innovative approaches such as parallel processing and model optimization techniques. Researchers are continuously exploring new architectures, like federated learning, which allows models to train across disparate data sources without necessitating data centralization.

Furthermore, the hardware aspect cannot be ignored. The von Neumann architecture, which has dominated computer design for decades, faces challenges with the data throughput demands of contemporary AI systems. Emerging hardware solutions, such as neuromorphic computing and quantum computing, offer promising directions. Neuromorphic computing mimics the human brain's neuronal pathways, potentially leading to a drastic reduction in power consumption and increase in processing speed. Meanwhile, quantum computing holds the promise of solving complex problems intractable by classical computers, especially those requiring simultaneous consideration of large datasets.

Distributed and cloud-based AI platforms are another cornerstone of scalability. As AI applications grow in complexity, they require distributed systems to handle computations simultaneously across various nodes. Cloud computing provides the infrastructure necessary for these operations, offering scalability in a cost-effective and efficient manner. Tech giants are investing heavily in cloud-based AI platforms that offer scalability as a service, empowering businesses to implement AI solutions without the need for massive capital investment in infrastructure.

Aside from technical infrastructure, ensuring the security, privacy, and ethical use of AI at scale is equally crucial. As systems grow, so do vulnerabilities and potential points of exploitation. Ensuring robust cybersecurity measures are in place becomes part of the scalability challenge. AI systems need to incorporate security mechanisms that evolve alongside them, protecting data integrity and privacy through encryption and other sophisticated methods.

The human factor is another layer of complexity in the scalability of AI systems. Scaling AI requires not just technical solutions but also skilled professionals who can develop, maintain, and oversee these expansive systems. The demand for AI expertise is growing rapidly, and educational institutions are being called upon to supply a workforce capable of understanding and advancing these technologies. This opens a broader conversation about education systems, training programs, and the global distribution of AI expertise.

As AI systems continue to scale, the interoperability between different systems and platforms becomes crucial. Scalable AI should seamlessly integrate with other technological frameworks and digital ecosystems without causing functional disruptions. Standards and protocols must evolve to facilitate this interoperability, ensuring smooth data flow and communication across diverse AI applications and industries.

Finally, while the technical aspects of scalability are immense, there is an inspiring dimension to this challenge. The drive to scale AI systems reflects a broader vision toward a future where intelligent technologies enhance human potential in ways previously unimaginable. By overcoming scalability challenges, AI can enable groundbreaking advancements in healthcare, education, environmental conservation, and beyond, creating a world where intelligent systems work in harmony with human endeavors.

The scalability of AI systems is therefore a multifaceted challenge requiring a convergence of novel technologies, robust infrastructure, and collaborative innovation. It's in tackling these challenges that AI can truly scale up and transform from a technological marvel to a practical tool that benefits society on a global scale. With continued focus and innovation, AI's scalability will become a cornerstone achievement in the ongoing narrative of human technological progress.

Overcoming Computational Limitations

The journey towards overcoming computational limitations in AI development is akin to navigating through both vast potential and significant obstacles. One of the most persistent challenges is the sheer amount of computational power required to train increasingly complex AI models. In recent years, models like GPT-3 and others in the transformer architecture family have demonstrated the immense potential of large-scale machine learning models. However, these groundbreaking developments come with a massive computational cost that can't be easily overlooked.

As the neural networks grow in complexity, they demand exponentially more processing power. This requirement outstrips what many traditional computing systems can offer. Enterprises and researchers are thus pushed to innovate, seeking novel ways to circumvent physical and financial barriers. Distributed computing and parallel processing are two paradigms that have emerged as crucial in this context. By distributing workloads across multiple nodes, they allow for the management of large datasets and complex model training, opening up new possibilities in AI development.

Furthermore, the advent of specialized hardware accelerators, such as graphics processing units (GPUs) and tensor processing units (TPUs), has been transformative. These high-performance units are designed specifically to handle the demands of AI computations, thus optimizing processing speed and efficiency. This hardware evolution has drastically reduced training times for sophisticated neural networks, allowing for more rapid experimentation and development. Still, the costs associated with using such hardware can be prohibitive, which limits accessibility for smaller research teams and startups.

An additional hurdle in overcoming computational limitations is the energy consumption associated with training and deploying AI models. High-performance computing tasks demand substantial energy resources, leading to increased operational costs and environmental impact. This creates a dual challenge of managing both the financial and ecological footprints of AI systems. Confronting this challenge has driven researchers to explore more energy-efficient algorithms and hardware, aiming to strike a balance between computational efficiency and sustainability.

Quantum computing presents an exciting frontier in this arena. While still largely in the experimental stage, quantum computing offers the promise of immense speed and processing capacity, far beyond the reach of classical computers. The potential for quantum computers to perform complex calculations almost instantaneously could revolutionize AI by dramatically reducing the limitations currently faced due to computational constraints. However, this remains primarily theoretical, with practical applications still on the horizon.

In tandem with advancements in hardware and computing paradigms, software optimization plays a critical role. Innovations in algorithm design can significantly reduce the computational load required for tasks without compromising on performance. For instance, techniques such as pruning, quantization, and knowledge distillation help streamline models, making them more efficient and viable for deployment on less powerful systems. These methods ensure that AI technology becomes more accessible and scalable, enabling its benefits to reach a broader audience.

Moreover, the rise of collaborative platforms and open-source initiatives has democratized access to advanced computational resources. Initiatives like cloud-based AI services provide scalable pay-as-you-go resources, allowing researchers and companies of all sizes to experiment and innovate without heavy upfront investment in infrastructure. Platforms like these offer both the computational muscle and the collaborative environment for diverse teams to share knowledge, data, and even incomplete models, accelerating the pace of breakthroughs.

The importance of interdisciplinary collaboration cannot be overstated. Addressing computational limitations in AI is not solely a job for computer scientists. It requires insights from electrical engineers, material scientists, and data scientists, among others. By merging expertise from various fields, innovative solutions can be crafted that tackle computational bottlenecks, engineer novel materials for better chip performance, and develop new methodologies to manage and process large datasets more efficiently.

Looking forward, the role of policy and regulation is becoming increasingly significant in managing the trajectory of computational advancements in AI. There needs to be balanced guidance to encourage innovation while ensuring that developments remain sustainable and accessible. Investments in education and training are also imperative to prepare a workforce capable of expanding and maintaining the new technological foundations being built.

Overcoming the computational limitations in AI development is a multifaceted endeavor that combines technological, environmental, and societal considerations. While the challenges are substantial, the ongoing innovations are inspiring and pave the way for a future where the constraints of today become the possibilities of tomorrow. The relentless pursuit of these solutions not only promises to advance AI technology further but also ensures it can be applied ethically and sustainably, enhancing industries and lives on a global scale.

Chapter 20:
User Experience Design in AI Agents

In a world swiftly becoming intertwined with AI, the design of user experiences in AI agents transcends mere interface aesthetics to forge profound human-centered engagements. Designing these experiences requires the delicate balancing act of making technology feel intuitive while ensuring it remains accessible to diverse populations. The key lies in crafting seamless interactions that anticipate user needs, empowering individuals and dismantling barriers historically posed by digital divides. As AI agents grow in ubiquity, focusing on accessibility and inclusivity not only enhances usability but also champions social equity, allowing more people to partake in this technological revolution. By fostering intuitive connections and emphasizing universal design principles, we pave the way for an inclusive future where AI serves as both guide and collaborator, amplifying human endeavors and enriching lives globally.

Creating Intuitive Interactions

As technology marches forward, the chasm between humans and machines steadily narrows. Central to this evolution is the ability of AI agents to facilitate intuitive interactions—those that require minimal effort from the user to understand and engage with. Designing these interactions involves a blend of psychology, design principles, and technological know-how. It's not about adding more features; it's about creating experiences that feel natural, human-like, and seamless.

Intuitive interaction design in AI is anchored in understanding human behavior. To predict and adapt to user needs, designers and engineers need to delve deep into cognitive psychology. The subtle art of anticipating a user's needs before they themselves might even know requires a level of personalization and empathy traditionally found in human relationships. This is where AI agents must excel; through continuous learning and adaptation, they can offer suggestions, recommendations, and solutions that align with the user's individual preferences.

Moreover, the success of an AI agent in providing intuitive interactions hinges on simplicity. Users should feel like they're interacting with an extension of themselves, rather than deciphering complex commands. This can be achieved by leveraging natural language processing, which allows users to communicate with AI agents in more familiar, conversational tones. By understanding natural language, AI agents can break down the barriers that traditionally impede human-machine interaction.

The interface through which users interact with AI is also a crucial component. Visual design plays an essential role in ensuring that AI-driven systems are both accessible and engaging. A well-designed interface should be aesthetically pleasing but, more importantly, it should prioritize usability. This means offering clear navigation paths, reducing cognitive load, and using intuitive imagery and icons. In essence, the visual aspects of interaction should guide the user effortlessly through the system without confusion or frustration.

Feedback mechanisms are pivotal in creating a responsive and interactive experience with AI agents. Users must receive timely and appropriate responses to their actions, whether it's acknowledging a command, reporting an error, or providing information. These feedback loops help in building trust and confidence. When users feel assured that the system has 'understood' them, they are more inclined to use it regularly and integrate it into their daily lives.

Furthermore, the adaptability of AI agents enhances intuitive interactions. An AI that learns from user interactions can refine its responses and suggestions over time. For instance, a digital assistant that observes your routine by recognizing morning habits can proactively suggest navigation routes or remind you of appointments. This adaptability ensures a dynamic and evolving interaction model catered specifically to the user's changing preferences and expectations.

The role of personalization in creating intuitive interactions cannot be overstated. By leveraging data aggregated from user behavior, preferences, and interactions, AI agents can tailor experiences uniquely suited to each individual. This moves the interaction from a one-size-fits-all model to a nuanced, personalized model. As AI systems become increasingly adept at understanding context and intent, they can present information and options that are most relevant to the individual user at any given moment.

However, the quest for intuitive interactions with AI agents also raises concerns regarding privacy. Personalization and adaptability necessitate access to vast amounts of data. Striking a balance between offering seamless interactions and safeguarding user privacy is a critical challenge. Developers must ensure transparency regarding data usage, providing controls to users and maintaining robust security protocols to protect sensitive information.

Accessibility is another critical factor when designing intuitive interactions. AI systems should cater to diverse needs, ensuring inclusivity so everyone, regardless of physical abilities, cognitive skills, or internet literacy, can interact effectively with AI agents. For example, voice-controlled systems offer a lifeline to those with limited mobility, while screen readers and alternative input methods support visually impaired users. AI-driven interfaces, when designed with accessibility in mind, can tear down barriers and extend their reach to the broader population.

By fostering an environment where AI agents are perceived as collaborators rather than tools, we both democratize technology and invite broader acceptance and trust. This approach encourages users to view AI as partners that enhance their capabilities rather than replace them. It crafts a narrative where human-centered AI becomes a medium of empowerment, offering support in navigating complex decisions and streamlining everyday tasks.

Ultimately, creating intuitive interactions with AI agents is an evolving field that challenges designers and technologists alike. It demands a harmonious fusion of technical prowess and empathetic design, geared to cultivate an ecosystem where users feel understood and valued. As AI continues to permeate various facets of life, its interaction paradigms must evolve from transactional to relational, fostering human-like synergies that are second nature.

Enhancing Accessibility and Inclusivity

As we move deeper into the era of artificial intelligence (AI), it's crucial to reflect on how we can design systems that are both accessible and inclusive. AI agents have the potential to reshape the world, but their impact will only be positive if they are designed with everyone in mind. Accessibility and inclusivity in AI agents mean crafting experiences that accommodate users of all abilities and backgrounds, ensuring no one is left behind.

Central to enhancing accessibility is the idea that AI agents must be usable by individuals with a wide range of abilities. This includes people with disabilities as well as those who may face temporary or situational impairments. For instance, voice-activated digital assistants can provide significant autonomy to users with visual impairments, allowing them to perform tasks hands-free. But it's not just about enabling basic functions—advanced design considerations can transform these interactions, making them effortless and empowering.

Moreover, inclusivity extends beyond physical accessibility to encompass cultural and linguistic diversity. AI agents should speak the languages of their users, understand regional dialects, and respect cultural nuances. An AI system operating worldwide must adapt its communication style to fit local idioms and sensitivities, thereby avoiding misunderstandings and fostering trust. This requires an underlying framework that supports multilingual capabilities and cultural intelligence.

A crucial component of achieving accessibility and inclusivity in AI is the concept of universal design. This approach aims to create products and environments that are inherently accessible, rather than requiring adaptations or specialized solutions. In this vein, AI developers can draw from lessons in architecture and education, where universal design has long been a guiding principle. The goal is to build systems that function seamlessly for everyone, without needing separate solutions for different groups.

Empathy is equally vital in designing AI experiences. Developers need to put themselves in the shoes of diverse users, considering how different people might interact with technology. Incorporating user feedback, especially from minority communities and people with disabilities, ensures that AI systems are intuitive and user-friendly. This participatory design approach not only creates better technology but also fosters a sense of ownership among users.

Advancements in natural language processing (NLP) offer promising avenues for making AI more accessible. By improving the ability of agents to understand and generate human language, we can break down barriers that might otherwise exclude non-native speakers or those with speech impediments. NLP can enable text simplification for those with cognitive disabilities or language-learning needs, providing explanations or instructions in straightforward terms without losing critical information.

One of the challenges in creating accessible AI systems is avoiding biases that can reinforce disadvantages. Machine learning models, which often drive AI agents, are prone to biases present in their training data. If not carefully managed, these biases can result in systems that discriminate against certain users or groups. Therefore, a critical part of inclusive design is ensuring representation in data sets and algorithms, which helps prevent the perpetuation of societal inequalities.

Beyond technical solutions, building an ecosystem that supports accessibility and inclusivity requires systemic change. Regulations and standards play an essential role in holding developers accountable. Organizations like the World Wide Web Consortium (W3C) have pioneered guidelines for accessible digital content, and similar standards are necessary for AI technologies. These guidelines serve as a baseline for innovation, encouraging developers to exceed minimum requirements and explore creative ways to include everyone.

The effort to enhance accessibility and inclusivity doesn't solely rest on the shoulders of developers; it requires a collaborative approach. Partnerships between tech companies, advocacy groups, and governmental bodies can drive forward best practices and innovations. By pooling knowledge and resources, such collaborations can better anticipate the diverse needs of users, addressing challenges before they become systemic issues.

AI holds the promise of leveling the playing field, making life easier for those with disabilities, or offering new opportunities for learning and connection. But realizing this vision requires dedication and foresight. As technological landscapes evolve, so too must our strategies for creating inclusive and accessible AI experiences. Embracing this challenge will not only ensure equity and usability but will enrich the field of AI, inspiring innovations that enhance lives globally.

Ultimately, the journey towards more accessible and inclusive AI agents is an ongoing one. It demands continuous reflection and adaptation as society, technology, and user expectations evolve. By committing to this path, we ensure AI can truly transform lives for the better, offering benefits that are universally shared rather than selectively distributed. Through thoughtful design and collaboration, AI agents can be powerful tools for inclusivity, championing a future that embraces every individual.

Chapter 21:
Social Impacts of AI Agents

Artificial intelligence agents, increasingly woven into the fabric of our daily lives, are reshaping societal dynamics in profound ways. As human reliance on these digital entities grows, our relationships with technology are evolving beyond mere interaction, fostering a deeper interdependence that challenges traditional norms. AI agents are altering how communities connect, communicate, and collaborate, compelling societies to redefine concepts of trust and reliance in the digital age. Simultaneously, the proliferation of these agents requires substantial adaptation, as individuals navigate the intricacies of coexistence with intelligent systems that influence numerous aspects of modern living. This transformation invites both opportunities for enhanced quality of life and implications for social equity, prompting dialogues about inclusivity and ethical integration across diverse cultures and communities. As we stand at the crossroads of human and machine collaboration, the societal ramifications of AI agents beckon us to contemplate not only their current impact but also their potential to reshape the very fabric of society in years to come.

Changes in Human-AI Relationships

As AI agents weave themselves into the fabric of our daily lives, the dynamic between humans and these intelligent systems continuously evolves. This shift in relationship reflects an intriguing convergence of technology and humanity, challenging our preconceptions of interaction, dependence, and trust.

For many, AI agents have become not just tools but entities that hold an almost human-like presence. Think about digital assistants like Siri and Alexa. They listen, respond, and even anticipate our needs, creating an illusion of companionship. These interactions, once mechanical, are increasingly natural and intuitive, subtly reshaping our communication habits. The evolving design of AI agents, with their enhanced ability to understand and emulate human emotions through sentiment analysis and natural language processing, blurs the line between machine capability and human-like understanding.

Yet, this burgeoning intimacy brings about several psychological and social implications. People might find comfort in confiding in AI without judgment or reprisal. The simplicity and convenience offered by these interactions can foster a sense of emotional attachment. However, this same comfort can lead to isolation from human contact, fundamentally altering how we associate with those around us. In some cases, individuals have reported preferring interactions with AI over humans due to the predictability and absence of conflict. This preference raises questions about the long-term impacts on social skills and human interaction.

The evolution of human-AI relationships prompts an observation on dependency. As AI agents handle more complex tasks—whether organizing our schedules, managing household operations, or even making critical business decisions—our reliance on them grows. This shift reallocates our mental resources, freeing us for more creative and strategic endeavors while potentially atrophying certain cognitive skills. But, there's a fine line between convenience and over-reliance, potentially leading to vulnerabilities if the technology fails or is tampered with.

Trust plays a pivotal role in this evolving relationship. For AI to be effectively integrated, there needs to be a level of confidence in its recommendations and decisions. Trust is built through transparency and understanding of how these systems work—a task easier said than done given the complexity of AI algorithms. As users, our inclination is to trust what we can see and understand; however, AI operates largely in the realm of the unseen, with complex models and data-driven decisions. The opaqueness of AI decision-making processes amplifies the challenge of engendering trust.

In recent years, the concept of anthropomorphism has gained traction in discussions about AI. As humans, we have an innate tendency to attribute human-like characteristics to non-human entities. AI agents, with their human-like voices, names, and personalities, often become the subject of this phenomenon. This personification can enhance user experience by making AI seem more relatable and engaging. However, it may also lead to unrealistic expectations regarding the capabilities and 'emotions' of AI, potentially resulting in disappointment or misplaced trust.

As we navigate this intricate relationship, ethical considerations come to the forefront. The responsible design of AI agents so they do not exploit human emotions or information is fundamental. Ensuring privacy and respecting user autonomy are critical components of ethical AI development. Furthermore, educating users about the limitations and actual capabilities of AI can help set realistic expectations and foster a healthier dynamic. Transparency in how AI systems function, combined with robust ethical guidelines, can guide the ethical adaptation of AI into the human social sphere.

Another aspect to consider is the adaptability and learning capability of AI agents. These systems not only execute pre-set tasks but also learn from every interaction. This continuous evolution allows for increasingly personalized user experiences, but it also raises

145

questions about identity and predictability. As AI becomes better at tailoring interactions based on user data, users might find themselves in echo chambers of their information and preferences, limiting exposure to new perspectives.

Interestingly, the transformative power of AI extends beyond typical human interaction into areas like leadership and decision-making. AI agents can now complement or even substitute human decision-making processes in certain contexts, providing data-driven insights that were previously inaccessible. This shift doesn't undermine human leadership but rather enhances it by enhancing decision-making with vast datasets and predictive analysis capabilities. However, this raises an important question: while AI can inform decisions, should it be allowed to craft policies or take on leadership roles where human intuition and empathy are indispensable?

For educators and policymakers, there's the challenge of preparing society for this evolving relationship. Developing new educational strategies to teach AI literacy, emphasizing the understanding of AI workings and implications, is vital. It's about equipping individuals not just to use AI, but to understand its impact and limitations to form balanced relationships with these powerful tools.

The unfolding changes in human-AI relationships display a fascinating narrative of technological symbiosis. As we welcome AI into more intimate segments of our lives, we're nudging towards a future that requires a recalibration of social norms and tech ethics. Embracing the potential is as important as navigating the associated risks and challenges, as our journey with AI isn't just about innovation but also about transforming the ways we connect, engage, and coexist with intelligent systems.

Societal Adaptation to AI Proliferation

As AI agents continue to proliferate across different sectors, societies worldwide are witnessing unprecedented changes. The integration of AI into daily life is akin to the advent of the internet or electricity—a transformative force reshaping how we live, work, and interact. This societal shift is not merely technological; it is cultural, economic, and philosophical. People are navigating unfamiliar landscapes where AI routines effortlessly blend with human activities.

One of the most palpable shifts is in how we define and interact with work. AI agents are automating routine tasks, providing people with more time to engage in creative and strategic endeavors. This doesn't just change job descriptions but is leading to the emergence of new career paths focused on AI oversight, maintenance, and augmentation. These changes require a dynamic approach to education, with a strong emphasis on continuous learning and adaptability.

Digital literacy is rapidly becoming as crucial as traditional literacy. Societies are increasingly emphasizing the need to include AI-related skills in educational curricula, starting from a young age. Schools and universities are rethinking their teaching strategies, striving to equip students with the necessary skills to thrive in an AI-enhanced world. The cultivation of skills like critical thinking, ethical reasoning, and complex problem-solving is prioritized, as these become the linchpins of human contribution in an AI-influenced society.

In workplaces, the line between human and machine work is fading. AI is acting as a collaborator in many fields, helping to make decisions and solve problems at scales that were previously unimaginable. In finance, healthcare, and even agriculture, AI agents collaborate with human experts to deliver precise and effective outcomes. Workers are leveraging AI tools to enhance productivity and creativity, rather than replacing the human touch. Companies and employees are navigating this landscape with a new mindset—one of partnership with technology.

The widespread adoption of AI also demands a deeper reflection on ethical values and human rights. With AI agents having access to vast amounts of personal data, societies are grappling with privacy issues and concerns over potential misuse. The prioritization of ethical AI development and implementation has become paramount. Regulations and policies are evolving, aiming to strike a balance between technological advancement and safeguarding human interests.

At the community level, AI is already playing a crucial role in empowering local initiatives. Smart city projects, driven by AI, are enhancing urban living by optimizing traffic flow, improving waste management, and increasing energy efficiency. These advancements help foster more livable urban environments, catering to the needs of growing populations while preserving natural resources.

AI's impact isn't limited to urban areas alone. In rural settings, AI agents are revolutionizing farming methods, contributing to sustainable agriculture practices, and improving crop yields. Access to AI-driven technologies in rural communities promotes equal opportunities and spurs economic growth, bridging the urban-rural divide. As AI continues to spread, communities are seeing improvements in communication, education access, and healthcare services, ultimately raising the overall quality of life.

Social interactions are also evolving with AI proliferation. The integration of AI agents in communication platforms is enabling more meaningful and inclusive interactions among individuals across the globe. Language barriers are being dismantled through real-time translation services, while AI-driven tools are aiding people with disabilities to interact more seamlessly within their environments. This fosters a more inclusive world, where technical advancements pave the way for shared understanding and empathy.

As societies adapt to AI, trust becomes essential. Earning and maintaining public trust in AI systems is crucial for widespread acceptance. Efforts to enhance transparency and explainability in AI processes are gaining traction, ensuring that users understand the mechanisms behind AI decisions and outcomes. The demand for AI systems that are fair, safe, and reliable emphasizes the importance of maintaining human oversight and accountability.

Education systems play an instrumental role in shaping perceptions of AI. By incorporating discussions of AI ethics, impact, and potential into the curriculum, societies are preparing future generations to navigate and shape the AI-laden world responsibly. Through informed discussions and inclusive policymaking, communities can effectively address societal concerns about AI and promote a balanced approach to its development and deployment.

As AI agents become ubiquitous, there is a cultural evolution underway. Society is redefining what it means to be human in a world where machines can perform tasks previously thought to require human intelligence. Philosophical discourse around AI is sparking new interpretations of creativity, consciousness, and agency, challenging pre-existing notions and fostering a deeper understanding of our own capabilities and limitations.

The art world, too, is experiencing a renaissance spurred by AI innovation. Artists are collaborating with AI agents to explore new forms of expression and creativity. Through these collaborations, societies are witnessing an amalgamation of human artistry and machine-generated content, creating hybrid art forms that push the boundaries of traditional aesthetics.

Ultimately, the societal adaptation to AI proliferation is a journey marked by challenges and opportunities. By embracing these changes, societies are fostering environments where AI can complement human achievements, enhancing quality of life and opening new avenues for progress. The future promises a symbiotic relationship between humans and machines, where collaboration and innovation drive collective growth and wellbeing.

In navigating this era of AI proliferation, societies must remember that technological advancements should serve humanity's broader goals. With thoughtful integration and management, AI has the potential to not only improve efficiency and productivity but to elevate the human experience. It's an era in which each societal sector plays a part, from government policymakers ensuring ethical guidelines to educators shaping AI-informed curricula, and industries pioneering innovative applications. Together, they can forge a future in which AI capabilities are harnessed for the greater good, leading to societal advancement and increased human flourishing.

Chapter 22:
AI Agents and Creativity

The realm of creativity, once considered uniquely human, is undergoing a radical transformation as AI agents become creative partners. These intelligent systems now assist artists in generating music, visual arts, and even literature, bringing new possibilities to artistic expression. AI agents are not merely tools; they're collaborators that challenge conventional notions of creativity itself. By blending computational efficiency with imaginative prowess, they help artists push the boundaries of their disciplines. This collaboration blurs the lines between human and machine creativity, posing profound questions about authorship, artistry, and the nature of creativity. As AI agents continue to evolve, they promise to unlock unprecedented creative potential, democratizing access to creative tools and widening the scope of what is perceived as art. This transformation offers a canvas as vast and varied as human imagination itself, encouraging us to redefine our creative landscapes in uncharted and exciting ways.

AI as a Tool for Artists

The realm of art has always been a dynamic landscape, shifting and evolving alongside technological advancements. With the advent of artificial intelligence, artists find themselves equipped with novel tools that expand the horizons of creativity and expression. AI acts as a powerful aide, providing not just new mediums but also new collaborators. This technological metamorphosis invites artists into a dialogue with machines, a partnership that challenges and redefines the boundaries of art itself.

Artificial intelligence offers artists an array of sophisticated algorithms that can mimic, learn from, and even innovate upon human creative processes. These tools range from neural networks that generate unique visual art to language models that assist in crafting compelling narratives. With AI, the artist's palette has broadened, making once-impossible feats accessible. For instance, generative art algorithms can swiftly produce intricate patterns and designs, which might have taken humans weeks or months to conceptualize and create.

Beyond merely mimicking human creativity, AI has the potential to lead artists toward uncharted territories. Artists can feed AI systems a mix of styles, textures, and compositions, allowing the system to extrapolate and produce hybrid works that blend various influences into a cohesive yet novel piece. This fusion not only broadens the scope of an artist's creative endeavors but also introduces audiences to pioneering styles that can be delightful surprises.

Generative models like GANs (Generative Adversarial Networks) and VAE (Variational Autoencoders) have become central to these artistic explorations. They enable the creation of surreal, dream-like artworks and transform simple inputs into complex masterpieces. Consider the case of AI in music – it can compose new pieces based on an existing music library, offering a fresh take on classical yet familiar sounds. This capability rekindles the old while simultaneously orchestrating the new, a dual task that is quintessentially artistic.

Moreover, collaboration between AI and human artists acts as a potent catalyst for conceptual thinking. AI can analyze vast datasets and suggest patterns and insights that a human might overlook, offering inspiration for new art forms or interpretations. For instance, in visual arts, AI can enhance an artist's perception by visualizing unseen patterns in shapes or colors, prompting reinterpretations of traditional concepts. Through this, AI becomes a source of inspiration, sparking creativity and innovation.

AI also plays a crucial role in democratizing art creation. By lowering the entry barriers to artistic ventures, AI allows individuals with minimal technical skills to express themselves creatively. Tools powered by AI can automate complex processes like 3D modeling or video editing, making the art creation process more accessible to beginners and experts alike. For many, this democratization means art is no longer confined to traditional practices or to those with formal training, instead becoming an inclusive space for all to explore.

Increasingly, AI is seen not as a replacement but as a partner in the creative process. This partnership encourages artists to use AI as a sounding board or creative partner in developing new artwork. By viewing AI as a tool, artists are empowered to push their boundaries without the fear of technological replacement. They are invited to experiment, to engage in dialogue with AI, and to harness its strengths to augment, rather than diminish, their own creative voices.

Furthermore, AI's role as a tool for artists goes beyond creation and extends into curatorial processes. AI systems can analyze and categorize vast collections of artworks, helping curators and collectors identify patterns or trends that might not be immediately apparent. This type of AI application is invaluable for institutions managing extensive art collections, paving the way for more dynamic exhibition experiences. By understanding these trends, stakeholders can also foresee shifts in cultural tastes, adapting their strategies accordingly.

Despite its advantages, the application of AI in art raises important ethical and technical questions. Who owns an AI-generated artwork? What are the moral implications of machines creating art? These questions hover in the background as the boundaries between human intention and machine output blur. Artists and technologists must navigate these waters thoughtfully, encouraging discourse around ownership, originality, and the essence of creativity itself.

As we push further into this age of digital innovation, artists are encouraged to embrace AI, exploring its potential to redefine creative processes. While AI tools can automate and analyze, the heart of the creative journey still resides with the human experience. It's our emotions, perspectives, and narratives that imbue art with meaning. AI is a tool, a co-creator in this journey, expanding what is possible without overshadowing the artist's individual vision.

In closing, AI has undeniably become an invaluable tool for artists. Its technologies, while rapidly advancing, are being integrated more seamlessly into artistic practices, providing enriched experiences both for the creators and audiences. As AI continues to evolve, so too will its interface with art, ensuring that creativity remains an ever-expanding universe filled with new possibilities and perspectives. Embracing AI doesn't mean surrendering to it but inviting it into a global tradition of artistic exploration and expression.

Blurring the Lines Between Human and Machine Creativity

The interplay between human and machine creativity is rapidly evolving as AI agents transform from computational tools into genuine collaborators in the creative process. In many domains, the distinct boundaries that once set humans apart from machines are becoming increasingly ambiguous. This chapter explores the intriguing fusion of human intuition with machine precision, illustrating how AI not only enhances creative endeavors but also challenges traditional notions of origination and authorship.

The integration of AI agents into creative workflows has sparked a renaissance of innovation. Artists, writers, and designers now find themselves collaborating with algorithms that have grown from completing basic tasks to offering insightful suggestions. These AI systems, trained on vast datasets, enable creatives to experiment beyond traditional limitations. This partnership pushes artists to explore novel ideas and techniques that might not have seemed feasible or even imaginable.

At the heart of this synergy is the concept of augmentation rather than replacement. AI agents serve to elevate human creativity by automating the repetitive tasks that can stifle the innovative spirit. This allows creatives to focus just on the aspects that require instinct and inspiration. For instance, in graphic design, AI can handle complex calculations and simulations, leaving the artist to concentrate on ideation and aesthetics. Far from being a replacement of talent, AI becomes a kind of muse, continuously sparking and refining human invention.

The world of music exemplifies this dynamic interaction. AI-generated compositions, once a novelty, are gaining recognition within the music industry. These algorithms analyze musical patterns and styles, creating melodies and harmonies that are fresh yet familiar. While AI can compose music autonomously, it often finds its greatest success in collaboration with human musicians. By providing material that artists can adapt and build upon, AI acts as a catalyst for new musical forms and expressions.

A similarly vibrant interplay is occurring in the literary field. Innovative authors and storytellers use AI to overcome writer's block, generate dialogue, or even develop entire narratives. These systems provide suggestions that can be quirky and unconventional, coaxing writers to explore paths they might not otherwise travel. However, the question of authorship looms large: Are such works collaborations or the fruits of machine labor? This raises philosophical debates about originality and creativity, challenging the notion of art as a purely human enterprise.

Despite these advancements, the fusion of human and machine creativity is not without challenges. One significant issue is the training data used to inform AI-generated outputs. Biases inherent in these datasets can propagate stereotypes or overlook certain creative styles, potentially marginalizing diverse voices and perspectives. As AI continues to play a significant role in creative decisions, it's crucial to curate diverse data that reflects a wider array of human experiences. Thus, diversity in data does not only promote inclusivity but enhances the richness of AI-generated creativity.

Moreover, as AI begins to contribute to cultural creations, questions about intellectual property rights become paramount. Who owns a piece of artwork created with AI assistance? Should the machine, its programmer, or its user claim authorship? Legal frameworks must evolve to address these questions, ensuring that creators are adequately protected while fostering an environment that encourages innovation and collaboration between humans and machines.

The potential of AI in artistic fields extends beyond traditional media. Technologies like generative design, where AI proposes multiple solutions to complex design challenges, are shaping new paradigms in architecture and industrial design. By iteratively refining prototypes, AI offers unique insights that human designers might overlook. This broadens the scope of what is creatively possible, effectively multiplying the creative potential of individual designers and teams.

In evaluating the influence of AI on creativity, it's crucial to acknowledge the symbiosis between technological advancement and human intuition. Machines may execute predefined algorithms to generate creative content, yet it is human empathy and depth of understanding that synthesize these pieces into something emotionally resonant. While AI can suggest countless color palettes or musical arrangements, only humans can discern which variant might evoke a particular emotion or tell a compelling story.

As AI continues to permeate artistic domains, cross-disciplinary approaches are burgeoning. Scientists and technologists collaborate with artists to push boundaries, resulting in hybridized projects that challenge conventional genres. By bridging the gap between technological and artistic disciplines, AI fosters an environment of open innovation, where creativity is seen as a universal language transcending human and machine capabilities.

Looking forward, AI's role in creativity will likely grow even more pronounced. Emerging technologies such as virtual reality and augmented reality, coupled with advanced AI, promise to unlock wholly new dimensions of human-machine interaction. These advancements will expand the canvas upon which creatives can innovate, crafting immersive experiences that blur the line between creator and audience, reality and imagination.

In summary, AI agents are more than tools—they are collaborators that amplify human creativity. By relieving humans of monotonous tasks and offering inventive suggestions, AI liberates the creative mind to soar to new heights. However, the journey to perfect synergy between humans and machines will require careful consideration of ethical, legal, and societal implications, encouraging a creative dialogue that embraces, rather than fears, the future. The ongoing dance between human intuition and machine precision is an exciting frontier, inviting us to rethink the very essence of creativity.

Chapter 23:
Future Trends in AI Agent Development

The development of AI agents is poised to accelerate, carving out pathways that seemed unimaginable a few years ago. As we stand on the brink of a technological revolution, AI is no longer confined to the realms of science fiction but is rapidly embedding itself in every facet of human life. In the near future, we expect AI agents to become even more sophisticated, capable of emotional intelligence, improved decision-making, and personalized interactions. The integration of quantum computing promises to exponentially increase processing power, enabling AI systems to tackle complex problems more efficiently. Meanwhile, advancements in machine learning and neural networks will push the boundaries of creativity and problem-solving, fostering innovation across industries. As AI agents become more autonomous, ethical considerations and responsible development will become paramount, demanding an ongoing dialogue to ensure these technologies benefit humanity as a whole. The future of AI agent development is not just about technological evolution but about redefining how we interact, work, and live, offering limitless potential and profound implications for society.

Emerging Innovations in AI Technology

Artificial Intelligence (AI) is relentlessly evolving with far-reaching implications and innovations emerging at an unprecedented pace. As AI technology becomes more pervasive in society, we're witnessing groundbreaking developments that are redefining what's possible. These evolving innovations form the very bedrock of future AI agent development, offering solutions that were once considered the realm of science fiction. At the heart of this transformation lies a fusion of creativity and machine intelligence, propelling us into an era of limitless potential.

One of the most captivating frontiers is the advent of swarm intelligence in AI systems. Inspired by nature, particularly the behaviors observed in ant colonies and flocks of birds, swarm intelligence leverages decentralized systems to solve complex tasks collaboratively. This approach shifts the focus from individual AI agents to collective intelligence, enhancing efficiency and robustness in problem-solving. These systems can tackle logistical challenges, optimize resource distribution, and even orchestrate large-scale, dynamic operations like disaster response.

Deep learning, a subset of machine learning, has also undergone significant leaps, challenging previous limitations with the help of technological innovations like neural architecture search (NAS). NAS automates the design of neural networks to maximize performance, allowing AI to essentially evolve its algorithms. This evolution leads to architectures that aren't merely designed by humans, exhibiting more adaptability and efficiency in tasks ranging from image recognition to natural language processing.

Another area witnessing substantial innovation is quantum computing, which holds transformative potential for AI. Quantum computers can process complex calculations at velocities unreachable by classical computers, providing a massive leap in the computational power available to AI systems. This advance could revolutionize fields requiring intense computational resources, such as cryptography, drug discovery, and extensive data analysis, enabling AI to solve problems currently beyond our grasp.

AI is also beginning to master emotional intelligence, a domain traditionally exclusive to humans. Emotion AI, or affective computing, seeks to recognize, interpret, and respond to human emotions. Through advancements in machine vision and audio analysis, AI agents can discern patterns in voice tones, facial expressions, and even physiological signals. These capabilities allow AI to offer more personalized interactions, enhancing customer experiences and creating emotionally resonant user interfaces.

Generative AI represents another phenomenal stride, empowering machines to create beyond simply analyzing and repeating existing data. Through generative models like GANs (Generative Adversarial Networks), AI can now craft art, write literature, compose music, and even design products. This creative partnership between humans and machines offers a glimpse into a future where AI fosters human creativity rather than replacing it, expanding the essence of artistic expression.

Connectivity is another vital component in the landscape of emerging AI technologies. The convergence of AI and the Internet of Things (IoT) is birthing intelligent ecosystems where everyday devices can communicate and collaborate. Smart cities exemplify this integration, utilizing AI and IoT to manage infrastructure efficiently, reduce energy consumption, and improve citizens' quality of life. This interconnectedness is fostering environments that adapt in real-time, optimizing everything from traffic flow to energy utilization.

Furthermore, the field of autonomous systems is achieving new milestones, particularly in mobility and robotics. Advanced AI algorithms are enabling autonomous vehicles to navigate complex environments with increasing precision. These vehicles are expected to revolutionize transportation, reducing accidents, and enhancing mobility for the disabled. In industries ranging from agriculture to manufacturing, autonomous robots are performing dangerous or mundane tasks, allowing humans to focus on more strategic roles.

The healthcare domain, too, is experiencing revolutionary AI innovations. From AI-driven diagnostic tools that offer early detection of diseases to predictive models that personalize treatment plans, healthcare is becoming more proactive than reactive. AI agents are aiding in drug discovery, speeding up the development of new medicines and therapies. By analyzing vast data sets from clinical trials and patient records, AI can uncover insights that might take decades to uncover through traditional research methods.

AI ethics and transparency have gained significant attention as well, with innovations focused on building trust in AI systems. Explainable AI (XAI) is becoming crucial, ensuring that AI processes are transparent and comprehensible to humans. This innovation addresses one of the most significant barriers to AI adoption, providing clarity on decision-making processes and ensuring that AI systems operate in fair, accountable, and non-discriminatory ways.

Cloud-native AI, an evolution within cloud computing and AI, is yet another forefront of innovation. Offering scalability, accessibility, and economic efficiency, cloud-native AI is democratizing access to powerful AI resources. Startups and large enterprises alike can deploy AI solutions without the need for extensive infrastructure, accelerating the cycle of innovation and making AI resources accessible to a broader range of users and applications.

AI's trajectory of innovation presents challenges and opportunities, shaping how society interacts with technology. In education, AI innovations are personalizing learning experiences, adapting to students' unique needs, and revolutionizing the educational landscape. In environmental management, AI is enhancing conservation efforts, optimizing resource utilization, and offering insights into complex ecological patterns, contributing significantly to the fight against climate change.

As with any transformative technology, responsible stewardship of AI innovations is critical. The balance between harnessing AI's potential and addressing ethical concerns will determine how these emerging innovations contribute to societal advancement. It's vital for forward-thinking policies and collaborations across sectors to guide the responsible evolution of AI technologies, ensuring that they serve humanity's best interests.

These innovations encapsulate the incredible potential AI holds to transform our world. As we continue to witness and shape these advancements, the goal remains to integrate AI seamlessly into our lives, enhancing human capabilities, and addressing global challenges. The future of AI is not a distant horizon but an unfolding reality, shaped by the continued interplay of innovation, ethics, and opportunity.

Predicting the Next Wave of AI Applications

As we gaze into the ever-evolving world of artificial intelligence, it's clear that the next wave of AI applications isn't just on the horizon; it's already making seismic shifts in how we interact with technology, businesses, and society. The nature of AI is such that it's shaped not only by technological advancements but also by the growing needs and imaginations of its users. We're at a pivotal moment where AI is not just supplementing human capabilities but enhancing them in unprecedented ways.

The next wave of AI applications is poised to transform industries by integrating more deeply into our daily lives. One of the most promising areas is natural interaction, where AI agents will break down the barriers between human and machine communication. With the advancement of natural language processing and machine learning, AI systems are learning to understand context, emotion, and intent much like humans do. This means that in the not-so-distant future, AI could offer empathetic customer support, real-time language translation, or even act as personal companions that engage in meaningful conversations.

In manufacturing and supply chains, AI is already optimizing processes and efficiencies, yet the future will see these systems becoming highly autonomous. Predictive analytics powered by AI will not just react to challenges but anticipate them. Imagine a factory floor where AI immediately adjusts production based on consumer demand forecasts, or supply chain systems that reroute shipments before disruptions can occur. This seamless integration of predictive models can drastically cut costs and reduce waste, offering economic and environmental benefits.

Healthcare, with its complexities and critical need for precision, stands as another frontier for AI evolution. The integration of AI for early diagnostics, personalized medicine, and automated data analysis is just the beginning. In the future, AI could potentially act as an always-available health advisor, leveraging real-time data from wearable technology to offer personalized health insights. By doing so, it might not only diagnose but also predict medical conditions before they manifest, considerably improving preventative care and patient outcomes.

We're also likely to witness AI's influence in creative fields, breaking the traditional boundaries of art, music, and literature creation. AI-driven algorithms are already generating awe-inspiring art and music, suggesting that these creative tools can partner with human creators to forge entirely new genres. Future AI applications will undoubtedly push this further, offering collaborative creation processes where humans and AI work hand in hand, redefining what art can be.

Moreover, the realm of transportation will see AI applications that handle more than just navigation. Autonomous vehicles could become deeply integrated into smart city ecosystems, interacting with traffic systems, pedestrians, and other vehicles in an intricate dance of data. This interconnectedness promises not only enhanced efficiency but also a reduction in accidents and emissions, revolutionizing urban planning and mobility.

Despite the promise, predicting the future of AI also entails recognizing the challenges. Ethical considerations, data privacy, and the need for robust regulatory frameworks must evolve alongside AI's capabilities. Ensuring that AI is inclusive, fair, and transparent remains a chief concern, one that demands a collaborative effort from technologists, policymakers, and users alike. Neglecting these facets can undermine the trust that is essential for AI's adoption and growth.

A major trend that will shape the future of AI applications is the convergence of AI with other technological advancements such as 5G, the Internet of Things (IoT), and quantum computing. This convergence will create synergies that amplify AI's capabilities beyond current limitations. For example, the fusion of AI with IoT can lead to hyper-connected environments, where data is processed and relayed instantly, enabling more dynamic AI responses in real time.

Furthermore, as quantum computing becomes more accessible, it may tackle some of the computational challenges faced by current AI systems. The ability to process vast amounts of data at unprecedented speeds could lead to breakthroughs in fields ranging from cryptography to pharmaceutical development, opening new avenues for AI applications.

AI's potential is boundless, but it will require nurturing through sustainable practices, continuous learning, and adaptive algorithms that evolve with changing human needs. Ecosystem-focused AI development, emphasizing sustainable growth and energy efficiency, will gain traction. Green AI, emphasizing reduced energy consumption and increased efficiency, will emerge as a guiding principle for future AI developments, contributing to meeting global sustainability goals.

In conclusion, the next wave of AI applications holds transformative potential for both industries and individuals. By harnessing emerging technologies and addressing core ethical and practical challenges, AI will continue to advance, becoming an ever more integral part of the tapestry of daily life. The future isn't just about creating smarter machines but fostering a more intelligent, collaborative, and sustainable interaction between humans and AI.

Chapter 24:
Preparing for a World with AI Agents

As we stand on the brink of an era dominated by AI agents, preparing for the transformative shifts they bring becomes crucial. The key to thriving in this new world lies in our ability to adapt, both individually and collectively, by embracing the changes in education and skill sets necessary for a future intertwined with AI. Our institutions must reimagine curricula to foster critical thinking and adaptability, equipping learners to work alongside intelligent systems. At the societal level, policies need to be crafted to ensure equitable access and address potential disruptions while promoting ethical AI development and deployment. This preparation isn't only about mitigating risks but also about seizing opportunities to harness AI for significant societal benefit. As these intelligent agents become more pervasive, our readiness will determine how they redefine the landscape of innovation and human potential.

Education and Skill Transformation

In a rapidly transforming world where AI agents are becoming ubiquitous, the way we learn and the skills we value are undergoing fundamental changes. Traditional education systems, designed for the industrial age, are scrambling to adapt to the demands of the digital era. This paradigm shift requires not just a superficial makeover but a deep rethinking of curricula, teaching methods, and the very goals of education.

AI's influence on education is twofold. On one hand, AI technologies offer innovative tools for personalized learning, making education more accessible and tailored to individual needs. On the other hand, AI poses a challenge: How do we prepare the workforce for roles that don't yet exist? While AI and automation may displace some jobs, they are also creating new opportunities that demand skills we might not have even considered essential a few years ago.

The first step in preparing for a world with AI agents is to foster a culture of lifelong learning. In the past, education was often seen as a phase of life—something that ended when one entered the workforce. Today, with technological advancements occurring at an unprecedented pace, a one-time education is no longer sufficient. The ability to continually learn and adapt is crucial. This doesn't just mean acquiring new technical skills; it also means cultivating critical thinking, creativity, and emotional intelligence—qualities that are harder to automate.

Moreover, interdisciplinary learning is gaining importance. The complex problems of the future will require insights and solutions that draw from multiple fields. Encouraging students to think across silos and appreciate the interconnectedness of knowledge can prepare them for this new reality. Schools and universities need to break down disciplinary barriers and promote curricula that integrate arts and sciences, technology and humanities.

Incorporating AI into the learning process itself can also be transformative. AI-powered tools like intelligent tutoring systems can offer personalized feedback, helping students identify their strengths and areas for improvement. These systems adapt to the learning pace of each student, fostering an environment where no one gets left behind. Similarly, AI-driven simulations and virtual environments can

167

facilitate experiential learning, allowing students to explore complex scenarios in a risk-free setting.

Teachers, too, will need to redefine their roles in this new landscape. Rather than being the primary source of information, educators will become facilitators of learning, guiding students as they navigate a wealth of resources and knowledge. Training programs for teachers should emphasize digital literacy and skills in managing AI-assisted learning environments.

At the policy level, governments and educational institutions must collaborate to develop frameworks that support these educational transformations. Investment in digital infrastructure is vital to ensure all students have equal access to the benefits of AI-enhanced learning. Additionally, policies should focus on aligning educational outcomes with the skills demanded by the labor market to avoid skill mismatches.

As AI continues to reshape industries, soft skills will gain prominence. Adaptability, ethical judgment, and teamwork are becoming as valuable as technical prowess. Employers are looking for individuals who can not only operate AI systems but also bring human insights to the table. Educational programs should aim to balance technical training with the development of these soft skills.

Furthermore, equity is a critical consideration in the transition toward AI-integrated education. There is a risk that increased reliance on technology will widen existing educational disparities. Ensuring equitable access to tech-infused learning environments is crucial for democratizing AI's benefits. Initiatives like providing resources to underfunded schools and offering community support programs can help bridge this gap.

Finally, the ethical implications of AI in education warrant attention. Questions related to data privacy, consent, and the transparency of AI algorithms used in learning environments need addressing. Students, parents, and educators should be involved in the discourse about how AI tools are used, ensuring that educational advancements don't come at the expense of personal privacy and ethics.

The journey toward education and skill transformation in an AI-dominated world is complex yet exciting. As AI continues to evolve, so too must our educational paradigms. The task is daunting, but by embracing the potential of AI and addressing its challenges head-on, we can equip current and future generations with the skills they need to thrive in any future.

Societal Shifts and Policy Integration

As AI agents become more integrated into the fabric of society, the impact on social structures is both profound and multifaceted. These agents are not merely technological tools; they represent a pivotal shift in how humans interact with machines and how societies organize themselves around this new axis. This transformation demands both individual adaptation and systemic changes at a societal level to harness the benefits while mitigating potential downsides.

The emergence of AI agents challenges traditional roles and occupations, prompting a re-evaluation of workforce dynamics. Jobs that primarily involve routine and repetitive tasks are particularly vulnerable to automation. However, this technological evolution is also a catalyst for the creation of new roles and opportunities that leverage human creativity, complex problem-solving, and emotional intelligence—traits where humans still surpass machines.

These changes necessitate a transformational shift in how societies approach education and vocational training. There's an urgent need to equip future generations with skills that complement AI efficiency, such as critical thinking, adaptability, and advanced technological competencies. Schools and educational institutions must embrace interdisciplinary learning that pairs technological fluency with the humanities, fostering a generation capable of ethical reasoning and innovative thinking.

In parallel, businesses and industries are compelled to integrate AI agents responsibly. The challenge lies in designing AI systems that align with human values and ethics, promoting transparency, fairness, and accountability. This is where the role of policy becomes crucial. Governments and regulatory bodies worldwide are tasked with crafting policies that ensure responsible AI deployment, balancing innovation with protection against potential misuse or harm.

The integration of AI agents into policy frameworks requires international collaboration and a consensus on ethical standards. Countries with diverse regulatory landscapes must work together to develop global norms for AI usage that uphold human rights and dignity. Such collaboration can lead to standardized guidelines that provide a clear framework for developers while ensuring AI is utilized for the common good.

Within local contexts, policymakers must consider the socio-economic impacts of AI agents on their communities. Policies should not only focus on technological advancement but also on fostering social welfare and equity. For example, AI-driven efficiency gains in public services can be reinvested in social programs that aid communities most impacted by automation-related disruptions.

Moreover, participatory governance models can play a critical role in shaping AI policy. Engaging diverse stakeholders—including technologists, ethicists, community leaders, and the public—ensures a holistic approach to AI integration. These dialogues can help crystallize what society envisions for its future with AI, balancing innovation with ethical considerations.

There's a growing recognition of the importance of integrating AI ethics into education and public discourse. By offering courses on AI ethics and digital literacy, societies can cultivate informed citizens who understand both the promises and perils of AI. Public awareness campaigns can help demystify AI technologies, dispelling myths and fostering an informed dialogue about their impact.

As we integrate AI agents into our lives, a fundamental societal shift in values and priorities becomes evident. There is a movement from a focus on consumption and efficiency toward sustainability and meaningful engagement. AI agents, with their ability to optimize resource use and predict outcomes, can be powerful allies in addressing global challenges like climate change and resource scarcity, provided they operate within a framework that prioritizes ecological and societal well-being.

Policy integration for AI also involves addressing concerns around data privacy and security. With AI systems requiring vast amounts of data to function effectively, robust data protection laws and practices are essential to safeguard individual privacy. Transparent data policies can build public trust in AI systems, ensuring that data is used responsibly and for legitimate purposes.

Finally, the social implications of AI agents extend to the very concept of community and civic interaction. As AI agents mediate more of our interactions, there's a potential risk of increased digital isolation. Policies and societal norms must emphasize the importance of maintaining human connections, encouraging interactions that nurture communal bonds and shared experiences.

In conclusion, the societal shifts brought about by AI agents are deeply interwoven with policy integration efforts. A collective, forward-thinking approach is essential to navigate these changes, ensuring they lead to a future that embodies both technological progress and the enrichment of human life. By embracing the transformative potential of AI with conscience and care, we can create a society where AI serves as both a tool for innovation and a catalyst for positive change.

Chapter 25:
AI Agents and Globalization

As the digital landscape transcends borders, AI agents are at the forefront of a transformative wave reshaping globalization. These agents are catalysts for significant shifts in global economies, driving efficiency and growth by automating routine tasks and unlocking new avenues for trade and inventory management. Their ability to process and analyze vast amounts of data facilitates cross-cultural exchanges, breaking language barriers through real-time translation and fostering collaboration that is unhindered by geographic or linguistic divides. The ripple effect of these advancements extends to diverse sectors, from supply chain optimization to inventive alliances between nations, spurring innovation and competitiveness on a global scale. As AI agents continue to evolve, they redefine not only how economies operate but also how societies interact and merge, heralding a future where the traditional boundaries of globalization are reimagined with unprecedented fluidity and interconnectedness, offering both opportunities and challenges that demand forward-thinking strategies and adaptability.

Impact on Global Economies

The transformative power of AI agents has begun to reshape the contours of global economies in unprecedented ways. These intelligent systems are more than just novel technological advancements; they are pivotal forces driving both economic growth and disruption. Their impact is pervasive, infiltrating sectors from finance to agriculture, altering the economic landscape on a global scale.

One of the clearest impacts AI agents have on global economies is through increased productivity. By automating routine tasks and optimizing complex processes, they have facilitated greater efficiency across industries. Manufacturing, for instance, has experienced a renaissance of sorts. Automated systems now handle everything from precision tasks in assembly lines to predictive maintenance in factories, resulting in faster production cycles and reduced downtime. This efficiency not only boosts the output of established markets but also lowers the barriers for emerging economies to step into global manufacturing roles.

Moreover, AI agents are powerful tools for innovation. They enhance research capabilities across various fields by providing insights that were previously unimaginable. In pharmaceuticals, AI has accelerated drug discovery processes, potentially saving billions in research and development costs. Such applications not only promise significant cost savings but also foster the development of new industries, creating jobs and stimulating economic growth across borders.

However, the deployment of AI agents also poses significant challenges. The skill gap between workers and the advanced technologies being adopted is widening. As machines take over more tasks, the demand for highly skilled workforces increases, leaving individuals in less advanced roles at risk. This shift could exacerbate income inequality both within and between countries, as economies struggle to recalibrate their workforces to meet new demands. Governments and organizations face the challenge of upskilling their populations to prepare them for this technologically advanced era.

The financial sector, in particular, has experienced a paradigm shift due to AI agents. Robotic trading systems process information and execute trades at speeds and efficiencies previously unattainable by human traders. While this results in more efficient markets, it also leads to the consolidation of wealth in fewer hands, as those with access to such technology hold a significant advantage. It's a high-stakes game where countries with the capability to develop and deploy AI systems may dominate the financial sector, leading to a reconfiguration of economic power globally.

Trade, too, is profoundly affected by AI. Intelligent algorithms can predict market trends, consumer behavior, and supply chain efficiencies, enabling companies to optimize their international trade strategies. This predictive power ensures that logistics and supply chains are streamlined, minimizing waste and maximizing profitability. It enhances trade efficiency but also requires countries to adapt their economic strategies to leverage these technologies effectively, ensuring they remain competitive.

On a geopolitical level, the proliferation of AI agents has significant repercussions. Countries investing heavily in AI technology bolster their international influence. This is not just about economic power; it's also about strategic advantage in defense, cybersecurity, and diplomatic negotiations. The race to dominate AI technology is becoming a critical component of national policy, prompting a reevaluation of international alliances and trade partnerships.

Interestingly, the economic impact of AI agents does not uniformly benefit all nations. Developing countries may find themselves at a disadvantage initially, due to a lack of resources and infrastructure to support widespread AI implementation. However, there's potential for these countries to leapfrog traditional development stages by integrating AI technologies into their economic frameworks. With strategic investments in AI education and infrastructure, they could transform into innovation hubs, driving regional economic growth.

It's worth noting that while AI agents offer numerous industrial and economic benefits, their impact also touches on more nuanced human aspects of economics. Consumer experience is increasingly personalized, leading to more efficient markets and better satisfaction. But as AI continues to integrate deeply into economic systems, there are concerns about data privacy, ethical governance, and the rights of consumers in an AI-driven market.

In response, governments worldwide are grappling with how best to regulate this new economic force. Policies that encourage innovation while protecting individual rights and promoting fair market competition are crucial. The establishment of global standards for AI implementation will require international cooperation, potentially redefining what economic collaboration looks like in the modern era.

Ultimately, AI agents stand at the heart of a transformative period in global economics. They bring about opportunities for unprecedented economic growth while posing challenges that require proactive management. As countries, industries, and individuals navigate this new economic reality, it becomes clear that AI agents will not only shape markets but also redefine the rules of engagement in our increasingly interconnected world.

Cross-cultural Exchange through AI

In the age of globalization, Artificial Intelligence (AI) agents are at the forefront of facilitating cross-cultural exchanges, acting as catalysts that transcend language barriers and foster deeper understanding among disparate cultures. With the increased interconnectedness of the world, AI's role in bridging cultural divides is more essential than ever before. The potential for AI to revolutionize how cultures interact lies in its ability to adapt and learn from vast datasets, extracting nuances that are often lost in human translation.

One of the most captivating aspects of AI's influence on cross-cultural exchange is its prowess in language translation. Traditional methods of translation, though diligent, often fail to capture the subtleties inherent in languages. AI, through advancements in Natural Language Processing (NLP), offers more dynamic and precise translations. Delightfully, tools like real-time translators and multilingual chatbots have emerged, making conversations between cultures almost seamless. These tools allow people to communicate and share ideas without the constraints of language barriers.

This technology is not just about words. It's about context, idioms, and cultural symbols that make language vibrant. AI can analyze and learn the cultural intricacies that come into play when people from different backgrounds communicate. By understanding the cultural context, AI can provide translations that are not just linguistically accurate but also culturally sensitive, ensuring the message maintains its intended meaning across borders.

Beyond language, AI agents also contribute immensely to cultural preservation and exchange. With the ability to process and analyze enormous amounts of cultural data, AI can assist in digitizing and preserving historical records, music, art, and folklore from around the world. As a result, cultural treasures that might have been lost to time are now accessible to anyone with an internet connection, fostering a greater appreciation and understanding of diverse cultural heritages.

Consider how AI is transforming the world of art and music. By examining different art forms and musical compositions from various cultures, AI agents can create new, hybrid art forms that blend elements from each. This convergence introduces people to previously unfamiliar styles and traditions, sparking interest and appreciation for what other cultures have to offer.

The educational sector is another realm where AI advances cross-cultural dialogue. AI-driven educational platforms have the potential to offer personalized, culturally contextualized learning experiences for students worldwide. Such platforms can tailor content not only to individual learning speeds but also to cultural preferences, promoting more inclusive and empathetic learning environments. As students learn about other cultures through AI-driven content, they gain a better understanding and appreciation for global diversity.

Moreover, AI agents are making significant strides in the area of cultural tourism. With virtual reality and AI, it's now possible to experience and learn about other countries and cultures from the comfort of one's home. These virtual experiences are designed to be immersive and educational, offering users a glimpse into the daily lives, traditions, and histories of people across the globe, thereby fostering empathy and understanding without the constraints of geography.

In the business world, AI is reshaping how companies approach cross-cultural collaborations. Through sophisticated data analytics, businesses can identify cultural trends and consumer preferences in different regions, creating products and marketing strategies that resonate with local audiences. This cultural insight is crucial for multinational companies aiming to maintain relevance in diverse markets, ultimately leading to more successful cross-cultural partnerships.

However, it's not all smooth sailing. There are challenges in ensuring that AI doesn't inadvertently reinforce cultural stereotypes. The risk of bias in AI systems is a pressing issue, as data used to train these systems may contain cultural assumptions or reflect a skewed view of certain populations. Hence, it's vital for developers and stakeholders to remain vigilant, actively working to identify and eliminate biases that could hinder genuine cultural exchange.

Further, while AI can indeed bring cultures together, there's also a pressing need for a human element. Genuine empathy and understanding require more than just data and algorithms; they require human connections. AI can facilitate these connections by providing the tools and platforms, but it's up to individuals and communities to use them with open minds and hearts.

To harness AI's full potential in cross-cultural exchanges, collaboration across borders is essential. Policymakers, developers, and cultural leaders must work together to develop AI systems that prioritize mutual respect and understanding. A cross-cultural lens should be a critical component in the development and deployment of AI tools, ensuring that they serve to unite rather than divide.

The future holds immense possibilities for AI as a mediator of cultural exchange. As AI technology continues to evolve, it will further enhance multicultural interactions, allowing for a richer and more profound understanding between people from all walks of life. The fusion of technology and culture, when approached thoughtfully, has the power to redefine how humanity perceives and interacts with the world.

In conclusion, AI agents are pivotal in shaping a future where cross-cultural exchanges are not just possible but are enriched. By bridging language gaps, preserving cultural heritage, and promoting global understanding, AI stands as a beacon of potential in bringing the world closer together. It reminds us that technology, when wielded with care and foresight, can be a powerful ally in the quest for a more interconnected and harmonious global society.

Conclusion

In traversing the myriad applications and implications of AI agents, we arrive at a crossroads that signifies not only the culmination of our exploration but also the beginning of a transformative journey for technology and society alike. The ascent of AI agents has been nothing short of remarkable, altering landscapes across industries and redefining what is possible through intelligent automation. This book has sought to demystify these complex entities, drawing connections between technological innovations and their real-world ramifications. As we stand on the precipice of a future characterized by profound AI proliferation, understanding these agents' roles becomes imperative.

AI agents have already left an indelible mark on sectors ranging from healthcare to retail, introducing efficiencies and capabilities that once existed only in the realm of imagination. These agents operate on the frontier of technology, utilizing machine learning, natural language processing, and data analysis to enhance human decision-making and streamline operations. Yet, as they integrate ever more deeply into the fabric of daily life, they also challenge us to rethink traditional notions of work, creativity, and human interaction.

Still, it is not all technological utopia. As AI agents become more pervasive, they undeniably bring ethical conundrums and privacy concerns to the forefront. The intrusive nature of intelligent systems demands robust discussions around governance frameworks and ethical guidelines that can balance innovation with individual rights.

Policymakers and tech developers must collaborate to create solutions that uphold societal values while fostering technological progress.

As we contemplate the future, it is crucial to prepare for the societal shifts AI agents inevitably provoke. Education systems must evolve to equip new generations with the skills necessary to thrive alongside advanced technologies. Beyond mere technical prowess, fostering critical thinking and ethical reasoning will empower individuals to use AI responsibly and innovatively. Meanwhile, businesses and governments will have to adapt rapidly to remain competitive in an AI-driven global market.

Moreover, AI agents present an unprecedented opportunity for cross-cultural engagement, offering a platform through which global economies can exchange ideas and innovations more freely. In doing so, these agents hold potential to bridge gaps between diverse communities, encouraging a shared technological destiny that respects cultural uniqueness while promoting global unity.

The future will also witness AI blurring the lines between art and science, with creativity emerging as a domain where human intuition and machine precision conspire to create groundbreaking works. As AI collaborates with artists, writers, and musicians, a new genre of creative expression is on the horizon, compelling us to reassess what constitutes originality and authorship.

Ultimately, AI agents herald a paradigm shift that echoes beyond technology—extending into the core of society where human values, ambitions, and philosophies reside. As such, the journey with AI is not just about leveraging its capabilities but aligning technological advancements with the essential tenets of human nature. This alignment calls for ongoing dialogue among scientists, ethicists, business leaders, and society at large, ensuring that the innovations of AI contribute positively to a shared future.

In looking forward, one finds a field teeming with potential yet fraught with uncertainty. The next wave of AI applications promises to redefine borders of knowledge, creativity, and interaction. It is a wave we cannot merely anticipate but must actively shape, guided by wisdom gleaned from past innovations and future aspirations.

The transformative power of AI agents embodies both challenge and opportunity. As we reflect on their journey so far and prepare for what's to come, we must remain vigilant stewards of this technology, ensuring its promise is realized for the collective advancement of society. The future beckons, and with thoughtful navigation, AI agents can indeed become a cornerstone upon which we build a more equitable, innovative, and prosperous world for generations to come.

Appendix A:
Appendix

In understanding the transformative landscape of AI agents, this appendix aims to provide additional insights and resources that complement the main text. While the previous chapters have explored various sectors and applications, this section serves as a coda, rounding out the discussion with further clarification on terms, methodologies, and studies pertinent to AI agents.

A comprehensive exploration of AI agents must take into account a myriad of factors, from the ethical and regulatory challenges to the promising opportunities posed by emerging technologies. Both the challenges and the potential have been underscored throughout the preceding chapters, setting the stage for innovative solutions that can shape the future.

In assembling this appendix, the intent is not only to serve as a glossary of sorts but to be a repository of additional data that may deepen your understanding. We recommend reviewing the key studies listed here, which are pivotal in illustrating how AI agents have evolved over time and continue to push the boundaries of what is possible.

Key Studies and Research Papers: Delve into pioneering works that have laid the groundwork for AI development, spotlighting projects and experiments that showcase the predictive capabilities and learning paradigms of AI systems.

Technical Jargon Explained: As a quick reference, this section demystifies the technical terms used throughout the text, so you can effortlessly navigate the complexities of AI jargon.

Additional Reading Recommendations: For those eager to explore beyond the fundamentals, curated lists of books, articles, and journals offer a pathway to expand your knowledge base.

This appendix is a testament to the multi-faceted nature of integrating AI agents across different domains. As technology continues to advance, the conversation surrounding AI will grow ever more complex, demanding continuous learning and adaptation.

While we have only scratched the surface of AI's potential in this book, the journey doesn't end here. This appendix is just one more step toward a future where AI agents are seamlessly woven into the fabric of our lives. Through continued exploration and application, we can harness the power of AI to solve today's challenges and tomorrow's mysteries.